MEN AMONG THE MAMMOTHS

SCIENCE AND ITS CONCEPTUAL FOUNDATIONS

David L. Hull, Editor

A. Bowdoin Van Riper

Men among the MAMMOTHS

. . . .

Victorian Science and the Discovery of Human Prehistory

. . . .

The
UNIVERSITY
of
CHICAGO PRESS

. . . .

*Chicago
and
London*

A. BOWDOIN VAN RIPER is visiting assistant
professor in the Science, Technology, and Society
Program, Franklin and Marshall College.

The University of Chicago Press, Chicago 60637
The University of Chicago Press, Ltd., London
© 1993 by The University of Chicago
All rights reserved. Published 1993
Printed in the United States of America
02 01 00 99 98 97 96 95 94 93   1 2 3 4 5
ISBN 0-226-84991-0 (cloth); ISBN 0-226-84992-9 (paper)

Library of Congress Cataloging-in-Publication Data

Van Riper, A. Bowdoin.
    Men among the mammoths : Victorian science and the discovery of
human prehistory / A. Bowdoin Van Riper.
        p.   cm.—(Science and its conceptual foundations)
    Revision of thesis (Ph.D.)—University of Wisconsin, 1990.
    Includes bibliographical references (p.   ) and index.
    1. Man, Prehistoric—Great Britain.   2. Archaeology—Great Britain—
History—19th century.   3. Anthropology, Prehistoric—Great Britain—
History—19th century.   4. Paleontology—Great Britain—History—
19th century.   5. Human evolution—Great Britain—Philosophy.   I. Title.
II. Series.
    GN805.v36   1993
    936.1—dc20   93–17006
                                                                           CIP

⊗   The paper used in this publication meets the
minimum requirements of the American National Standard
for Information Sciences—Permanence of Paper for
Printed Library Materials, ANSI Z39.48–1984.

CONTENTS

FIGURES

ACKNOWLEDGMENTS

THIS BOOK BEGAN, in January 1987, as a page-long idea written on a single sheet of paper. It has grown a great deal since then, both in scope and in bulk, but hardly by my efforts alone.

The staffs of the Torquay Natural History Museum (in Torquay, England) and the Falconer Museum (in Forres, Scotland) offered generous hospitality as well as generous access to their collections. The staffs of the Edinburgh University Library, the British Library, and the libraries of the American Philosophical Society, Royal Society of London, and Geological Society of London guided me to a wide variety of published and unpublished materials. I am grateful to the Museums Division of the Moray District Council, the Edinburgh University Library, and the Torquay Natural History Society for permission to quote from unpublished materials in their possession.

Closer to home, I am indebted to libraries and librarians too numerous to list. The libraries of my home institutions—the University of Wisconsin at Madison, Northwestern University, and Franklin and Marshall College—deserve special mention, however. Their staffs performed bibliographic miracles on a regular basis, locating all but the most obscure nineteenth-century texts and often providing multiple editions. I owe a particular debt to John Neu and Marie Dvorzak at Wisconsin, and to the unsung heroes who oversee the microforms and interlibrary loan departments of all three institutions.

When it was still a dissertation, the text benefited in countless ways from the comments and advice of my committee: Victor Hilts, Lynn Nyhart, and Margaret Schabas of the History of Science Department, Robert Dott of the Geology and Geophysics Department, and James Stoltman of the Anthropology Department. Since then, many others have been equally generous with their time and expertise. Colleagues at the University of Wisconsin—Madison, Northwestern University, and Franklin and Marshall College have all listened to my ideas and asked thoughtful, incisive questions. Countless friends outside academia have also contributed, asking, "What's your book *about*, anyway?" and, by asking, encouraging me to define the answer more clearly. I am indebted, most of all, to David Hull and two anonymous referees for reading the entire manuscript and offering detailed, chapter-by-chapter suggestions for improving it.

Susan Abrams, along with her staff at the University of Chicago Press, has given me all the help and enthusiastic support that an author could want from an editor. Her patient responses to my questions and concerns demystified, at every stage, the process of turning a manuscript into a finished book.

Fellowships from the Wisconsin Alumni Research Foundation and Northwestern University's Program in the History and Philosophy of Science allowed me the luxury of writing (and then rewriting) full time during the 1989–90 and 1990–91 academic years. Franklin and Marshall College has given me other luxuries in 1992–93: the flexible scheduling, superb computer system, and supportive working environment that have allowed me to finish the book while I began my teaching career. My parents have offered support in many forms, of which financial assistance was only one.

Finally, my thanks go to Julie Newell. She has provided dozens of bibliographic references, listened to countless explanations of my ideas, and read thousands of pages of manuscript. Through it all, she has paid me the compliment of being my most demanding editor. Her eye for flimsy arguments and murky prose has been unblinking, and her tireless efforts have made this book far better than it could ever have been without her. It is my great good fortune to have her for a colleague, a friend, and my wife.

A NOTE ON TERMINOLOGY

I HAVE TRIED to avoid using technical terms except where they reflect distinctions that are essential to the story being told. The word "mammoth," for example, functions throughout the book as a generic term for two species of extinct elephant—*Elephas antiquus* and *Elephas primigenius*. Victorian geologists distinguished between the two species but regarded them as coeval and thus (for chronological purposes) interchangeable. The phrase "chipped-stone tools" covers a wide range of artifacts produced, in different styles and for various purposes, by essentially similar processes.

Readers familiar with the geologic time scale should note that Victorian geology's "Post-Pliocene" and "Recent" epochs correspond to modern geology's "Pleistocene" and "Holocene" epochs. "Pleistocene" first appeared in the 1830s, but confusion over its precise meaning led British geologists to reject it in favor of "Post-Pliocene." I have, for reasons of internal consistency, referred to the two epochs by their Victorian names. "Post-Pliocene," as used in this book, thus denotes a discrete geological epoch and not the entire span of time after the end of the Pliocene Epoch.

All measurements are given, again for reasons of consistency, in English rather than metric units.

Monkeyana

Am I satyr or man?
Pray tell me who can,
And settle my place in the scale.
A man in ape's shape,
An anthropoid ape,
Or monkey deprived of his tail?

The *Vestiges* taught,
That all came from naught
By "development," so-called "progressive";
That insects and worms
Assume higher forms
By modification excessive.

Then DARWIN set forth,
In a book of much worth,
The importance of "Nature's selection";
How the struggle for life
Is a laudable strife,
And results in "specific distinction."

Let pigeons and doves
Select their own loves,
And grant them a million of ages,
Then doubtless you'll find
They've altered their kind,
And changed into prophets and sages.

LEONARD HORNER relates
That Biblical dates
The age of the world cannot trace;
That Bible tradition,
By Nile's deposition
Is put to the right about face.

Then there's PENGELLY
Who next will tell ye
That he and his colleagues of late
Find celts and shaped stones
Mixed up with cave bones
Of contemporaneous date.

Then PRESTWICH, he pelts
With hammers and celts
All who do not believe his relation,
That the tools he exhumes
From gravelly tombs
Date before the Mosaic creation.

Then HUXLEY and OWEN,
With rivalry glowing,
With pen and ink rush to the scratch;
'Tis Brain *versus* Brain,
Till one of them's slain;
By Jove! It will be a good match!

Says OWEN, you can see
The brain of Chimpanzee
Is always exceedingly small,
With the hindermost "horn"
Of extremity shorn,
And no "Hippocampus" at all.

The Professor then tells 'em,
That man's "cerebellum,"
From a vertical point you can't see;
That each "convolution"
Contains a solution,
Of "Archencephalic" degree.

Then apes have no nose,
And thumbs for great toes,
And a pelvis both narrow and slight;
They can't stand upright,
Unless to show fight,
With "DU CHALLIU," that chivalrous knight!

Next HUXLEY replies,
That OWEN he lies,
And garbles his Latin quotation;
That his facts are not new,
His mistakes not a few,
Detrimental to his reputation.

"To twice slay the slain,"
By dint of the Brain,
(Thus HUXLEY concludes his review)
Is but labour in vain,
Unproductive of gain,
And so I shall bid you "Adieu!"

GORILLA
Zoological Gardens, May 1861
                    Punch, May 18, 1861

ONE

. . . .

Beyond the Present World

PALEOANTHROPOLOGY may be the last science with a genuinely romantic reputation. The study of modern humans' prehistoric ancestors is linked, in the public mind, with an image as familiar as a scene from a favorite old movie: A lone khaki-clad scientist walks, head down, across a barren landscape somewhere in East Africa. Presently, the scientist stops, kneels and begins to scrape away at the dusty soil. A smooth, dark, rounded object gradually emerges. The scientist digs deeper, exposing eye sockets and teeth. Finally, the ancient skull is fully exposed. The scientist, gently freeing it from the ground, grins in triumph—humankind's understanding of its own origins will now be a little clearer.

Other aspects of the science—the ones involving laboratories, museums, and papers delivered at conferences—have been chronicled in detail by both participants and observers.[1] Debates over alternate models of human evolution have received more press coverage than most recent scientific controversies. Yet, in the public mind, paleoanthropology remains the preserve of lone scientist-explorers who make crucial discoveries in exotic locales.

Paleoanthropologists would be the first to point out that neither their fieldwork nor their field lives up to the public's romantic image. The search for modern humans' earliest ancestors is a

1. Readable histories of paleoanthropology include J. Reader, *Missing Links,* 2d ed. (1988); R. Lewin, *Bones of Contention* (1988); and D. Willis, *The Hominid Gang* (1990). P. J. Bowler (*Theories of Human Evolution* [1986]) and M. Landau (*Narratives of Human Evolution* [1991]) treat the theoretical side of the discipline but focus on the late nineteenth and early twentieth centuries.

1

cooperative endeavor. A modern expedition in search of early hominids typically includes not only paleoanthropologists but also paleontologists, archaeologists, geologists, and a corps of local collectors familiar with the area and its fossils. Success—the discovery of hominid remains—would draw in additional specialists including geochemists to assign a date to the bones and anatomists to help reassemble them. The diversity of such expeditions reflects their purpose: to reconstruct both the early hominids and the world that they inhabited. It also reflects the assumption that early hominids, their tools, and their environment cannot be understood in isolation from one another.

The study of our oldest ancestors is thus interdisciplinary both in principle and in practice. What is striking about this state of affairs is not that it exists today but that it is already more than a century old. It reaches back to the mid-nineteenth century and predates, by decades, paleoanthropology's existence as a distinct discipline. Cooperative, interdisciplinary studies of early humans and their world began in Europe during the late 1860s. The studies, and the newly redrawn disciplinary boundaries that made them possible, were products of a new answer to a venerable scientific question: "How old is the human race?"

THE RECENT ORIGIN of the human race was a central element of Christian cosmologies for more than 1,500 years. Implied by scripture, it was validated by the work of seventeenth-century biblical scholars such as Archbishop James Ussher, who calculated that the Earth was 6,000 years old and the human race only days younger.[2] It was central to Thomas Burnet's *Sacred Theory of the Earth* (1681–84) and similar attempts to provide scientific explanations for the major events in Earth history. Burnet's *Sacred Theory*, and subsequent works with similar titles and aims, presented Earth history and human history as virtually coeval. The authors' explanations of events such as the Deluge drew on the testimony of ancient texts, both secular and sacred, as well as observations of present-day rocks and fossils.

2. W. R. Brice, "Bishop Ussher, John Lightfoot and the Age of Creation," *Journal of Geological Education* 30 (1982): 18–24.

The age of the Earth grew by orders of magnitude in the work of eighteenth-century natural philosophers such as Benoît de Maillet and the Comte de Buffon, but human history remained within its traditional 6,000-year limits. As Earth history grew, therefore, human history shrank in proportion. James Hutton, writing a century after Burnet, described an Earth so old that all but an insignificant fraction of its history had passed without human witnesses to observe it. Human recency thus came to have a twofold meaning during the eighteenth century. The human race was young in terms of years but also young in comparison to the Earth and its vast array of plant and animal life. Geology, archaeology, and ethnology each incorporated the twofold idea of human recency as they separated, during the late eighteenth century, into intellectually distinct disciplines. The idea that humans had appeared 6,000 years ago, in the last moment of Earth's long history, shaped the new sciences' boundaries and the terms on which their theoretical battles were fought.

During the early nineteenth century, British geologists systematically excluded the origin and early history of man from their studies of Earth history—a decision that, ironically, made the age of the human race an important geological question. The origin of man, like the origin of the Earth itself, also had significant religious implications. Leading geologists, hoping to insulate their new science from religious controversies and promote cooperation between geologists of all denominations, therefore "fenced off" both the earliest and latest portions of Earth history. Geology, they argued, would empirically investigate the history of the Earth and the history of life in the long intervening period. Geology's intellectual domain encompassed the "former worlds" populated by now-extinct animals, but excluded the "modern world" inhabited by man. Geology ended, and archaeology took over, at the point where human remains began to appear alongside plant and animal fossils.[3]

Ideas about the age of the human race shaped archaeology's

3. R. Porter, "Creation and Credence," in *Natural Order,* ed. B. Barnes and S. Shapin (1980); M. J. S. Rudwick, "The Shape and Meaning of Earth History," in *God and Nature,* ed. D. Lindberg and R. Numbers (1986). C. C. Albritton (*The Abyss of Time* [1980]) discusses changing ideas about the earth's antiquity.

intellectual content as well as its temporal boundaries. The brevity of human history, and the large portions of it documented by written records, implied that civilization had developed quickly. Ancient Egyptian writings suggested that the complex civilization of the pharaohs was already established by 2000 B.C. Sven Nilsson and J. J. A. Worsaae—Danish archaeologists who noted that the pre-Christian inhabitants of Europe had used first stone, then bronze, and finally iron tools—argued that bronze and iron artifacts were introduced by technologically sophisticated invaders.[4] The alternative explanation, that bronze- and ironworking were local innovations, implied a slower and more gradual process than humankind's brief history allowed.

The image of successive invasions (or, more benignly, migrations) sweeping across Europe and Asia also appeared in contemporary ethnological works. Monogenists, who argued that all human races had descended from a single set of ancestors, used such migrations to account for the rapid dispersal of humans to even the most remote areas. To explain the physical differences between races, they postulated that early humans had been more susceptible than modern humans to environmental influences. Neither postulate rested on a strong foundation of evidence, but the brevity of human history made both vital to the monogenists' case. Polygenists, who argued that each race had descended from a separate set of ancestors, used the recent origin of the human race as a basis for their own theories and a means of attacking their opponents.' They pointed to 4,000-year-old Egyptian art depicting Arabs and Ethiopians with recognizably modern facial features. How, they asked their monogenist rivals, could the races have taken on their modern form only 2,000 years after humans first appeared on Earth?

Belief in the recent origin of the human race was as much a part of early Victorian religious thought as it was of early Victorian science. It provided one of the last points where science's emerging picture of Earth history reinforced traditional interpre-

4. For a brief treatment of Nilsson and Worsaae see G. Daniel, *150 Years of Archaeology* (1975), 38–54, and B. G. Trigger, *A History of Archaeological Thought* (1990), 73–86. For the social and intellectual context of their work, see O. Klindt-Jensen, *A History of Scandinavian Archaeology* (1975).

tations of Genesis, but its influence on religious ideas about "man's place in nature" was even more significant. Human recency reinforced what many Christians saw as a central message of Genesis: the idea that humans were the greatest of God's creations and lords of the planet that He had created for them.

Early nineteenth-century divines believed, as Thomas Burnet had, that the presence of humans on Earth gave meaning and purpose to Earth history. They abandoned Burnet's assumption that Earth history and human history were coeval but embraced a similar idea: that humans had appeared on Earth shortly after the advent of the "modern world." Geologists began, in the late eighteenth century, to break the long history of life on Earth into distinct periods, each of which possessed a unique population of plants and animals perfectly adapted to prevailing conditions. The more imaginative saw each period as a "former world"—an alien landscape populated by strange beasts that had long since become extinct.[5] Mammoths, giant cave bears, and wooly rhinoceroses had roamed the plains of a bitterly cold Europe during the last of these former worlds. That world had passed away—to be replaced by the familiar climate, plants, and animals of the modern world—shortly before the advent of the first humans.

The idea that humankind and the modern world were nearly coeval was important to both scientists and clergymen. It was interpreted as evidence that God had created the modern world specifically for humans, and it encouraged both groups to seek evidence of Him in the details of natural history. Works such as William Buckland's *Geology and Mineralogy Considered with Reference to Natural Theology* cataloged the ways in which God had designed the modern world to meet human needs.[6] Linking humans to the recently formed modern world also gave theological significance the Earth's long prehuman history. The linkage implied that man, the crowning glory of God's creation, did not appear on Earth until the Earth had been prepared for him—that

5. N. A. Rupke, *The Great Chain of History* (1983), 109–49.

6. C. C. Gillispie, *Genesis and Geology* (1951), is the classic study of this subject. More recent works include J. H. Brooke, "The Natural Theology of Geologists," in *Images of the Earth*, ed. L. Jordanova and R. Porter (1979), 39–62; J. H. Brooke, *Science and Religion* (1991), 192–225; and Rupke, *Great Chain of History*, 231–54.

is, until the familiar features of the modern world were in place. The entire prehuman history of the Earth, apparently marked by geological activity of diminishing violence and life of increasing complexity, could be interpreted as a long, gradual process of preparation designed to create a world suitable for humans. Though chronologically insignificant, the few thousand years of human history thus gave meaning and purpose to the untold eons that preceded them.

The Victorian religious community, committed to the belief that both nature and revelation offered evidence of God, embraced these ideas. Devout scientists employed them in essays, textbooks, lectures, and other writings that lay outside the ostensibly theory-free world of formal scientific papers. The educated public, too, was familiar with the implications of human recency. The recent origin of the human race, and the scientific and theological significance of that origin, remained a central element in Victorian science until the early 1860s. It formed part of a common intellectual context that allowed clergymen and lay people, as well as scientists, to discuss ideas about the past.[7]

JOHN FRERE was an English gentleman with a modest political career and a taste for the emerging sciences of archaeology and geology. History has little else to say about him, and his reasons for visiting the small Suffolk village of Hoxne in the spring of 1797 have long since been forgotten. The only record of his trip is a brief letter that he wrote to the London Society of Antiquaries reporting an extraordinary discovery.

Frere's letter described a pair of primitive tools, chipped from flint, that he had found in a local brick-earth quarry. The artifacts, Frere stated, were "evidently weapons of war, fabricated and used by a people who had not the use of metals" (fig. 1). He did not regard them as "objects of curiosity in themselves;" archaeologists had recognized such objects as primitive weapons for more than a century.[8] What made weapons from Hoxne unusual, Frere argued, was the circumstances in which they had been found (fig.

7. R. M. Young, "Natural Theology, Victorian Periodicals, and the Fragmentation of a Common Context," in *Darwin's Metaphor* (1985).

8. D. K. Grayson, *The Establishment of Human Antiquity* (1983), 6–8.

Figure 1. One of the chipped-stone tools that John Frere discovered at Hoxne in 1797 (from Frere, "Account of Flint Weapons," 1800).

2). They came from a bed of gravelly soil, ten feet below the surface, that held dozens of similar artifacts. A layer of sand—above the gravel bed and thus deposited more recently—contained marine mollusc shells and the bones of an enormous animal unlike any alive today. The depth at which the tools were buried, and the presence of marine shells and an extinct animal above them, suggested their great age. Such circumstances, Frere concluded,

Figure 2. Cross-section of the deposits at Hoxne, as they existed at the time of Frere's 1797 visit. The gravelly soil contained stone tools and bones of extinct mammals, while the layer of sand above it contained marine shells (based on Frere, "Account of Flint Weapons," 1800).

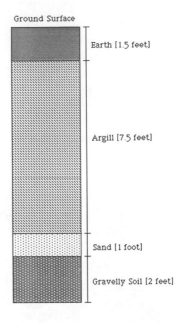

Ground Surface

Earth [1.5 feet]

Argill [7.5 feet]

Sand [1 foot]

Gravelly Soil [2 feet]

"may tempt us to refer them to a very remote period indeed; even beyond that of the present world."[9]

The London Society of Antiquaries, to which Frere addressed his letter, was Britain's leading archaeological society. The letter was read before its members in June 1797, published in the 1800 edition of its journal, and forgotten. To an archaeological community primarily interested in Greek and Roman artifacts, the primitive weapons of ancient Britons were, indeed, not "objects of curiosity." Geologists and ethnologists, who might have been interested in Frere's implications about the age of the human race, also paid little attention. They possessed, in 1800, neither national societies of their own nor clearly defined research programs. Geologists' knowledge of Europe's rocks and fossils remained too sparse to provide a meaningful context for Frere's observations. The report from Hoxne thus remained an isolated piece of information, and was soon lost amid the dozens of other brief reports that filled the British archaeological literature.

In 1859, sixty-two years after Frere discovered them, the

9. J. Frere, "Account of Flint Weapons Discovered at Hoxne in Suffolk," *Arch.* 13 (1800): 204–5.

stone tools from Hoxne came before the Society of Antiquaries for the second time. John Evans, a respected archaeologist, read a long paper on primitive stone tools at the society's June 2 meeting. Describing artifacts he had found in France the month before, he cited Frere's letter as proof that "discoveries of precisely similar weapons and implements had been made under precisely similar circumstances in this country." Joseph Prestwich, a respected geologist who had accompanied Evans to France, had made a similar announcement to the Royal Society of London the week before.[10] Evans and Prestwich were, like Frere, gentleman scientists familiar with both geology and archaeology. They described observations and drew conclusions that echoed Frere's, noting that stone tools—primitive, but clearly the work of human hands—had been found in gravel beds of great antiquity. Evans and Prestwich did not speak to an apathetic world, however, and by 1859 the Hoxne tools were not an isolated datum. Evans and Prestwich's 1859 revival of Frere's observations was part of a concerted attempt to challenge long-standing ideas about the past.

Evans's and Prestwich's papers were only two among a half-dozen that were read before Britain's leading scientific societies between March and September 1859. The papers discussed different data, but each came to the same conclusion: humans had lived alongside now-extinct mammoths and rhinoceroses at a time when Western Europe was cold enough to support herds of reindeer. The authors' estimates of the exact age of the human race differed, but all agreed that it was greater than the long-accepted figure of 6,000 years. The data underlying their conclusions came from sites in England, France, and Sicily; they were supplemented, during the next three years, by fresh data from dozens of additional sites. Charles Lyell, Britain's leading geological theorist, summed up the new evidence for human antiquity in his *Geological Evidences of the Antiquity of Man* (1863). Lyell's book brought the case for "men among the mammoths" before the educated public and gave it the stamp of scientific orthodoxy.

10. John Evans, "On the Occurrence of Flint Implements," *Arch.* 38 (1860): 298. J. Prestwich, "On the Occurrence of Flint-implements," *Philosophical Transactions of the Royal Society* 150 (1860): 304–8.

The papers that Evans and his colleagues presented in 1859 drew immediate skeptical responses from other scientists, to which advocates of human antiquity responded both in person and in print. The debate broadened, after the publication of Lyell's *Antiquity of Man,* to include members of the clergy and the educated public. The controversy over human antiquity—like the controversy over Darwinism that quickly overshadowed it—revolved around scientific discoveries, but it was fueled by concerns about the implications of those discoveries. The new case for human antiquity, like Darwin's case for evolution by natural selection, challenged long-cherished ideas about humankind's place in nature.

NEW SCIENTIFIC ideas are created within an intellectual context that they, in turn, reshape. Neither their genesis nor their impact can adequately be understood without close attention to that context. Nor can the context be narrowly defined. New theories have the potential to alter not only the content of the discipline that produced them but also the relationships between disciplines—to change what David Allen felicitously refers to as the intellectual topography.[11] The case for human antiquity that John Evans announced in 1859 was a product of the same intellectual milieu as Charles Darwin's theory of evolution. The 1859 papers, like Darwin's *Origin of Species,* changed the way that educated Victorians thought about Earth history and humankind's place in it. Nineteenth-century scientists' ideas on the human antiquity question have been studied in detail. Unlike Darwin's work, however, they have never been firmly placed in their intellectual context.

Donald Grayson's *The Establishment of Human Antiquity* (1983) surveys sixty years of work on the subject by British and Continental scientists. It ranges from John Frere in 1797 to John Evans in the 1860s, paying close attention to the logic of both proponents' and opponents' ideas on human antiquity. Grayson demolishes the myth that the evidence for human antiquity was sufficient to convince any unbiased observer, and that opposition

11. D. E. Allen, "The Lost Limb," in *Images of the Earth,* ed. L. J. Jordanova and R. Porter (1979).

to the idea must therefore have been the result of scientific or religious dogmatism. His close analysis of evidence and argument shows that, before 1859, there was good reason to doubt the claims of human antiquity theorists such as Jacques Boucher de Perthes. The success of Prestwich, Evans, and their colleagues in 1859 depended, Grayson shows, as much on their excellent scientific reputations as on the quality of their data.

A small group of shorter works on the establishment of human antiquity serves as a useful adjunct to Grayson's survey. Jacob Gruber's "Brixham Cave and the Antiquity of Man" (1965) examines the work that revived British interest in the human antiquity question in 1858. John Lyon's "The Search for Prehistoric Man" (1970) profiles five early, unsuccessful advocates of human antiquity and examines the religious implications of finding men among the mammoths. W. F. Bynum's "Charles Lyell's *Antiquity of Man* and its Critics" (1983) focuses on contemporary charges that the book was derivative and Lyell's acknowledgment of others' work too sparse. Finally, several authors have discussed the impact of the 1859 case for human antiquity on archaeological and anthropological theory.[12]

These works provide a clear outline of the discoveries and published papers that established human antiquity, but they also raise intriguing questions. Why did British geologists, whose interest in the human antiquity problem had remained desultory for forty years, suddenly begin an intensive effort to resolve it in 1858? Why did British archaeologists take so little interest in the age of the human race, despite its (retrospectively) obvious significance for their science? What effect did the dominance of geologists in the human antiquity research of the late 1850s and early 1860s have on that research? When, and how, did the search for the earliest humans become a cooperative venture involving archaeologists and anthropologists, was well as geologists?

12. L. K. Clark, *Pioneers of Prehistory in England;* Daniel, *150 Years;* K. P. Oakley, "The Problem of Man's Antiquity," *Bulletin of the British Museum (Natural History), Geology Series* 9 (1964): 86–153; Bowler, *Theories of Human Evolution;* P. J. Bowler, *The Invention of Progress* (1989); T. R. Trautman, *Lewis Henry Morgan* (1987); and F. Spencer, *Piltdown* (1990).

THIS IS the story of the establishment of human antiquity from a fresh perspective: it integrates a detailed narrative of how Victorian scientists raised and resolved the human antiquity problem with a survey of the problem's broader intellectual context. The narrative rests, methodologically, on the work of Martin Rudwick, James Secord, and David Oldroyd, who have demonstrated the importance of ideas exchanged in the pages of letters and behind closed committee-room doors.[13] It proceeds from the idea that, throughout the Victorian era, new geological theories and interpretations were more often framed in private by groups of specialists than in the field by individuals. It draws, therefore, on unpublished as well as published correspondence, and on field notes as well as polished scientific papers.

Evans, Prestwich, and their colleagues first presented their case for human antiquity in the spring of 1859—at the beginning of a period of intellectual crisis. The intellectual context within which Victorians discussed the human antiquity question was, as a result, very broad. Arguments about whether there had been men among the mammoths shaped, and were shaped by, controversies over Darwinian evolution, the proper interpretation of the Bible, and the relationship between God, nature, and humankind. Human antiquity affected not only the boundaries and day-to-day practice of geology and archaeology but also solutions to the larger problem that T. H. Huxley dubbed "Man's Place in Nature."

LIKE ALL HISTORIES of the human antiquity question, this book revolves around the events of 1858–63. The six principal chapters fall neatly into three pairs that might be labeled "Before," "During," and "After." The first pair of chapters surveys the intellectual topography of early Victorian science and examines the relationship of the human antiquity problem to archaeology and geology in the 1840s and 1850s. The second pair analyzes the five-year period (1858–63) in which a half-dozen geologists created, defended, and popularized a new consensus on the age of

---

13. M. J. S. Rudwick, *The Great Devonian Controversy* (1985); J. A. Secord, *Controversy in Victorian Geology* (1986); D. R. Oldroyd, *The Highlands Controversy* (1989).

the human race. The third pair of chapters traces, on two distinct levels, the impact of the new consensus after 1858–59. Chapter 6 examines the commentaries on the new consensus published by clergymen, lay people, and scientists outside the "core set" responsible for creating and expanding it. Chapter 7 analyzes the several, interrelated ways in which the establishment of human antiquity shaped the intellectual topography of later Victorian science.

The search for evidence of the earliest humans has always been an international enterprise. British, American, French, Belgian, Swiss, Danish, and German scientists were all involved during the sixty-odd years between John Frere's discovery at Hoxne and the announcement of a new consensus in London. A few words about the geographic limits of this study—its focus on British scientists and their ideas—are, therefore, in order.

The evidence on which the 1859 case for human antiquity rested came principally from sites in Britain and France. The scientists most deeply involved with it—and later with studying the beginning of the European Stone Age—were also British and French. German, Swiss, and Scandinavian scientists, confronted with other types of primitive remains, focused most of their attention on later periods of prehistory.[14] The new consensus on human antiquity that emerged between 1858 and 1863 was widely discussed in both Britain and France, but its origins were clearly British. Its architects were all British geologists, and investigations behind it were rooted in a research program unique to British geology. Britain also offered, particularly after 1860, unique stimuli to debates over human antiquity and it implications. The publication of Darwin's *Origin of Species* and the belated arrival of hermeneutic methods from Germany clashed with the long-standing penchant of the British for blending science and theology and created a hothouse atmosphere not present in France.

This is not to say that the history of the human antiquity problem begins and ends in Britain. The story of French scientists' work on the problem, and reaction to its solution, deserves a ful-

14. B. G. Trigger, *A History of Archaeological Thought* (1990), 108–9. There were, of course, exceptions to this admittedly broad generalization.

ler treatment than it has yet received. The story of human antiquity research in North America prior to 1890 also deserves fuller
treatment. Americans saw the human antiquity question in archaeological, rather than geological, terms. They tied it, moreover, to the politically sensitive question of whether the continent's first inhabitants had been Indians or a more advanced,
now-vanished race known as the Moundbuilders.[15]

THE SCIENTIST'S ostensible goal is to subsume a complex set
of phenomena under a comparatively simple law. It is ironic,
then, that the historian of science frequently seems bent on dis-
mantling simple tales of discovery and erecting long, complex
narratives in their place. I have, in the pages that follow, tried to
do something similar to the establishment of human antiquity.
The idea that there had been men among the mammoths did not
arise, fully formed, from a careful study of stone tools and fossil
bones. Sixty years of objections and doubts were not wiped away
overnight by exhortations to go, look at the evidence with unbiased eyes, and be convinced. The redrawing of boundaries between disciplines and the blurring of old boundaries between
present and former worlds did not—could not—take place in an
intellectual vacuum.

   The establishment of human antiquity was the product of a
unique intellectual world that it, in time, changed beyond recognition. It produced ideas about the past, and ways of understanding the past, that are still in use today. That knowledge frames
the questions this book addresses, but does not dictate the answers. What follows is an attempt to present the establishment of
human antiquity as nineteenth-century participants and specta-
tors saw it: tied to contemporary scientific problems, complicated
by personal and institutional rivalries, and laden with both religious and philosophical implications.

   15. D. J. Meltzer, "The Antiquity of Man and the Development of American
Archaeology," *Advances in Archaeological Method and Theory* 6 (1983): 1–51; R. Silverberg, *Mound Builders of Ancient America* (1968).

TWO

The Historical Archaeologists and Their Science

IN 1851, the year that the Crystal Palace opened and tens of thousands flocked to London to gape at the wonders of the modern age, Britons were also fascinated by their island's distant past. The rapid proliferation of archaeological societies reflected this growing interest in the past and its material remains. Of the two national and forty-one county archaeological societies founded in England between 1830 and 1880, half were established during the fifteen years between 1840 and 1855.[1] The new societies made archaeology a nationwide enterprise for the first time in Britain. By the late 1850s, it was an established scientific discipline with a large, enthusiastic community of researchers and a clearly defined research program. Three national journals and a dozen or more local ones gave archaeologists a forum for their ideas, and the annual summer meetings of the British Archaeological Association and the Archaeological Institute drew large crowds.

British archaeologists of the 1840s and 1850s shared, and at the national level explicitly prescribed, a radically new approach to the past. Their insistence on scientific rigor and their goal of reconstructing the past in detail set them apart from earlier archaeologists. Their belief that archaeology and history were close allies distinguished them from the prehistoric archaeologists who became a distinct community in the early 1860s. The archaeologists of the 1840s and 1850s coined no specific name for themselves or the set of methods and goals that they shared.[2] The

1. For a complete listing see P. J. A. Levine, *The Amateur and the Professional* (1986), 182–83.
2. The founders of the new approach to archaeology used both the traditional term *antiquaries* and the newer term *archaeologists* to refer to themselves.

terms used in this book—*historical archaeologists* and *historical archaeology*—are designed to distinguish their research program from that of the later prehistoric archaeologists. Historical archaeology represented the mainstream of British archaeological thought from the mid-1840s through the early 1860s. It was within the context of historical archaeology that British archaeologists evaluated the evidence that Prestwich, Evans, and others displayed as proof of human antiquity.

The heyday of historical archaeology was brief. Born in the mid-1840s, it was displaced from the cutting edge of archaeology in the late 1860s. Though still popular, it was less prominent and influential than the new research program—focused on prehistory and allied with natural science—that John Evans, John Lubbock, and others founded (see chapter 7). Historical archaeology has received little attention from historians, who have dismissed its practitioners as unscientific at best and destructive nuisances at worst.[3] This characterization is neither charitable nor accurate. A sense of common purpose and a shared, clearly articulated body of methods and techniques united the historical archaeologists. Their approach to the past, though substantially different than that of later archaeologists (including those of the present day), was no less scientific. The structure of the historical archaeologists' research program reflected their view of the past and of their proper role as students of that past. The same intellectual commitments shaped their lack of interest in the human antiquity problem and, in turn, their reactions to the crucial discoveries made in 1858–59.

A Community and Its Beginnings

The Society of Antiquaries of London, founded in 1717, dominated British archaeology for the next century and a quarter. It

---

Antiquary and *antiquarianism* did not yet carry their modern connotation of casual, unscientific dabbling and an obsession with trivial details.

3. The classic statement of this view is G. Daniel, *150 Years of Archaeology* (1975), 79–84. Similar claims appear in G. Daniel, *The Idea of Prehistory* (1962), 23–25; B. G. Trigger, *A History of Archaeological Thought* (1990), 71–72; and Levine, *Amateur and Professional*, 13–23. All four works mistakenly equate scientific archaeology with a focus on prehistoric artifacts, the extensive use of excavation, and acceptance of the Three-Age System for subdividing prehistory.

began as a modest association of twenty-three men, many of them members of the Royal Society, who felt that the older body did not take sufficient interest in "the valuable Relicks of former Ages."[4] The first pages of the society's first minute book, dated New Year's Day 1717–18, outline its founders' intentions. The society's founders began by asserting that "the Study of Antiquitys has ever been esteem'd a considerable part of good literature," and argues for the importance of preserving "the venerable remains of our ancestors," as well as "what will assist us in a clearer Understanding of the invaluable Writings of the Antient Learned Nations." The work of the society's members, they continued, will be to "collect and print and keep exact Registers under proper Heads Titles of all Antient Monuments that come into their hands whether Ecclesiastic or Civil, which may be communicated to them from all parts of the kingdoms of Great Britain and Ireland, such as . . ." A list of thirty-three items, intended to be suggestive rather than definitive, followed. Its scope was vast, taking in every imaginable relic of Britain's past from "old cities," castles, and temples to maps, charters, and genealogies.[5]

The Society of Antiquaries began as a small organization with a clearly defined purpose, but it did not remain so for very long. Its membership rose to 173 by 1764, and ten years later was approaching 300. Despite this growth, the society remained socially homogeneous: its members were those who had sufficient leisure time to pursue their study of the past, and who lived either in London or close enough to make at least occasional visits possible. These implicit "membership requirements"—along with the expense of initiation fees and annual subscriptions—meant that most members were wealthy, titled, or both. After 1770, stipulations that each new member must be sponsored by three existing members and approved by a two-thirds majority made it easier for the society to regulate the "quality" of the membership, and ensured the dominance of wealthy Londoners.[6]

Perhaps because of its largely upper-class membership, perhaps because of its richly decorated quarters in central London, the Society of Antiquaries became a fixture on the London social

4. Joan Evans, *History of the Society of Antiquaries* (1956), 58.
5. Ibid., 58.
6. Ibid., 148–49 ff.

scene in the late eighteenth century. By 1784, it consisted of 376 Fellows, one of whom remarked in a letter to a friend that it had "become one of our most fashionable weekly rendès vous's. Instead of old square toes you now behold smooth faces and dainty thin shoes with ponderous buckles on them."[7] This new fashionability may have added to the conviviality of society's meetings, but it did nothing for the quality of the papers read there. As the society's reputation as a place to be seen rose, its intellectual reputation gradually eroded. In 1788, an anonymous writer in the *Gentleman's Magazine* complained that "he who looks into the *Archaeologia* for profound researches into the ancient history, laws, poetry or manners, of Britain, will be entirely disappointed; and will find the whole eight volumes to contain only amusing fugitive papers, on ancient buildings, monuments, medals, etc. with a few indeed of more importance intermixed."[8]

The society's problem, as the critic in the *Gentleman's Magazine* realized, was that its members were more collectors of antiques than students of the past. They saw the antiquities they acquired as art objects, to be valued for their beauty or uniqueness, rather than as relics of a vanished civilization, to be valued for the light they could shed on the lives of those who made and used them. For most members of the Society of Antiquaries, archaeology was an acquisitive, not an intellectual, enterprise. A collection of antiquities was as much a part of a well-furnished upper-class home as were a large library and a cabinet of natural curiosities.[9]

A mid-eighteenth-century revival of interest in ancient Greek culture, combined with an educational system firmly rooted in Greek and Roman texts, exacerbated the problem. Interest in antiquities from Greece, Italy, and the Near East—once confined largely to the classically oriented Society of Dillettanti— became fashionable within the Society of Antiquaries as well. A steady stream of upper-class Englishmen headed for the Mediterranean, determined to study the past and bring pieces of it home

7. James Douglas to Brian Faussett, February 4, 1875, quoted in Joan Evans, *Society of Antiquaries*, 187.

8. *Gentleman's Magazine* 58 (1788), pt. 2, suppl., 1149, quoted in Joan Evans, *Society of Antiquaries*, 197.

9. M. Girouard, *Life in the English Country House* (1978), 173–74, 178.

to display. Once they reached foreign soil, their interest was piqued by the discovery that Mediterranean artifacts were more attractive and more readily obtainable than British ones. The glory of Greece and the grandeur of Rome—sculpture, vases, and pieces of buildings—could be bought for a pittance or carried off for free.[10] Lord Elgin, who shipped the Parthenon's marble friezes home to the British Museum, was merely the best-known figure in a well-established tradition.

There were exceptions to the eighteenth-century archaeologists' view of artifacts as art objects. The most famous of them, as well as the most prolific, was William Stukeley. Stukeley firmly believed that the study of Britain's antiquities should be practiced in the field. He carried out extensive, single-handed surveys in the south of England during the first half of the century, drawing and describing a wide variety of ancient monuments. His interpretations grew increasingly fanciful toward the end of his career, but the standards of his descriptive work were uniformly high. Stukeley's books on Stonehenge and the even larger earth-and-stone structure at Avebury are unique records of how the two monuments looked before the cumulative results of erosion and indiscriminate land-clearing operations damaged them.[11]

Stukeley was one of the first British scholars to systematically excavate burial mounds and other ancient earthworks in an attempt to understand them. The tradition was carried on by the Reverend Bryan Faussett in the 1760s and 1770s and by Captain James Douglas of the Royal Army Corps of Engineers in the 1780s and 1790s. Sir Richard Colt Hoare and his associate, William Cunnington based their two-volume *Ancient History of Wiltshire* (1812–19) on field surveys and excavations. They are often cited—along with others who did similar work—as the "grandfathers" of modern archaeology, and there are good reasons for bestowing such a title.[12] They recognized the potential value of

---

10. R. Jenkyns, *The Victorians and Ancient Greece* (1980), 1–20; Daniel, *150 Years*, 20–21.

11. S. Piggott, *William Stukeley* (1985); Piggott's *Ancient Britons and the Antiquarian Imagination* (1989) relates Stukeley's work to that of other eighteenth-century antiquaries.

12. See, e.g., Daniel, *150 Years*, 29–33; and B. Marsden, *Pioneers of Prehistory* (1984), 5–24. For further biographical material on William Cunnington, see R. H.

excavation, kept careful records of their work, and used methods that (considerably modified) are still in use today. It is important, however, not to overestimate their influence.

Stukeley, Douglas, Cunnington, and other individuals like them were just that: individuals. They were neither typical members of the antiquarian community nor members of a distinct subcommunity of their own. Separated from each other in both time and space, they had little contact and no opportunity for sustained discussion or cooperative research. Even the famous team of Cunnington and Colt Hoare operated independently when in the field; Colt Hoare would survey a region on horseback and draw up a list of barrows for Cunnington, who took charge of nearly all the digging.[13] The absence of a truly national archaeological society, or any other accessible forum for publicizing new discoveries, intensified the isolation of individual fieldworkers—particularly those who lived far from London.

As a result, the pioneering work of Stukeley and his successors never coalesced into a coherent research tradition. Cunnington was almost certainly aware of Stukeley's work, but his own studies were not the intellectual descendents of Stukeley's. Faussett, Douglas, and their lesser-known contemporaries may have known each other, but they were not bound together by allegiance to a particular method or belief in a particular view of the past. The field-workers of the eighteenth and early nineteenth centuries thus exerted only a limited influence on antiquarian theory and practice.[14] The business of collecting, displaying, and describing portable bits of the past went on—for most antiquaries—as it had always done.

The Society of Antiquaries itself had, as late as the mid-1840s, changed very little. It was still large, still a fashionable rendezvous, and still plagued by papers that were nothing more than excruciatingly detailed descriptions of some object recently added

Cunnington, *From Antiquary to Archaeologist* (1975); for Richard Colt Hoare, see K. Woodbridge, *Landscape and Antiquity* (1970), 187–234; for James Douglas, see R. Jessup, *Man of Many Talents* (1975).

13. Marsden, *Pioneers of Prehistory*, 17.

14. For an illuminating contrast, see R. Porter, *The Making of Geology* (1980), 16–30.

to the author's collection. One prospective member, after attending a meeting in 1839, reported that he "came away not vastly impressed with the talents of the body assembled. . . . All these men, no not all, but the powers that be are vastly behind the time. It is quite comical to see how they copy and borrow from each other how thoroughly conventional they are, what little freshness of mind they possess."[15] The charges leveled against the society's journal by Nicholas Harris Nicolas in 1830 remained all too valid. "[Its] volumes consist chiefly of lucubrations on broken stones, potsherds, tumuli, and runic inscriptions, or of interminable essays on armor, relieved now and then by a letter from the Museum, which the secretary, driven, to adopt his usual words, 'by the dearth of other communications,' is forced to hunt for at the last hour."[16]

Incompetent leadership compounded the problems created by the shortage of worthwhile papers. The president, George Gordon, earl of Aberdeen, was so preoccupied with his political career that he did not attend a single meeting in the last six years of his term. The treasurer remained oblivious to the details of the society's account book and to the fact that twenty-five members were more than ten years behind in their dues. The secretary, despite a respectable salary, would do nothing for the society without being paid extra for it and took a cavalier approach to publishing the papers that he received. One member complained in 1839 that two papers he had submitted in 1812 had still not appeared in print. The council, consisting of the society's nine officers together with members appointed by the officers and approved by the president, was hardly inclined to push for reform.[17]

In the mid-1840s, the insularity and intellectual petrification of the Society of Antiquaries produced two nearly simultaneous reactions. Middle-class enthusiasts, new to the study of archaeology and deeply interested in the history of the provincial towns where they lived, founded the first county archaeological socie-

15. Charles Hartshorne to Albert Way, April 30, 1839, quoted in Levine, *Amateur and Professional*, 51.
16. Nicholas Harris Nicolas, *Observations on the State of Historical Literature* (1830), 23, quoted in Joan Evans, *Society of Antiquaries*, 249.
17. Joan Evans, *Society of Antiquaries*, 240–51.

ties. As local societies began to blossom in southern and eastern Britain, six disenchanted members of the Society of Antiquaries met in London to found a new national society, the British Archaeological Association. Within a year, internal political squabbles caused two of the association's founders to secede and form the Archaeological Institute, but it was both intellectually and organizationally similar to the association.[18] The local and national societies were founded by men from different intellectual backgrounds and pursued somewhat different agendas. The members of both soon became integrated into a single community, however. Their goals were broadly similar, and their memberships overlapped significantly.

The new county archaeological societies emerged during a period of rapid social and intellectual change in England. Four of these changes—the rise of the middle class, the romantic movement, the Oxford movement within the Church of England, and the growth of railways—were particularly significant for the development of archaeology. No single factor was primarily responsible for the growth of the county societies.[19] All four played significant roles in expanding the appeal and accessibility of the expanding discipline.

The expansion of the middle class significantly broadened the demographic base of the archaeological community. During the eighteenth and early nineteenth centuries, the heavy demands archaeology made on leisure time and disposable income had limited its participants to the titled and the wealthy. By the 1840s, the growth of interest in nearby British antiquities and the advent of cheap railway travel had made these demands less onerous, and the number of people capable of meeting them had also increased. Moreover, the new members of the middle class were resolutely determined to waste neither time nor money on

18. E. R. Taylor, "The Humours of Archaeology," *JBAA*, n. s., 38 (1932): 183–234; Joan Evans, "The Royal Archaeological Institute," *AJ* 56 (1949): 1–5.

19. For the influence of the Romantic Movement see S. Piggott, "Prehistory and the Romantic Movement," *Antiquity* 11 (1937): 31–38; and Daniel, *150 Years*, 20–21. For the Oxford Movement see S. Piggott, "Origins of the English County Archaeological Societies," in *Ruins in a Landscape* (1978). For the rise of the middle class, see Trigger, *History of Archaeological Thought*, 74–85. For the railways see K. Hudson, *A Social History of Archaeology* (1981), 43–49.

idleness and frivolity. This desire for self-improvement and "rational amusement," which helped to fuel the natural history craze of 1820–80, performed a similar social function for archaeology.[20] As natural history could turn a seaside excursion into an educational experience, so archaeology could ennoble a picnic in the shade of picturesque ruins.

The opportunities for "rational amusement" that archaeology offered were only part of its attraction. It appealed to the emotions as well, satisfying the middle-class public's taste for knowledge of "Britain's glorious past" and of its more spectacular material relics: old parish churches, ruined abbeys, and crumbling castles. Patriotic pride and nationalism accounted for some of the growing interest in British antiquities. Stuart Piggott has shown, however, that the growing popularity of archaeology can also be traced to the cultural echoes of the romantic movement and to the later, church-based Oxford movement. Together, the two movements gave British ruins an air of grandeur and majesty that had long been associated only with classical ruins.

The romantic movement encouraged a new, aesthetic appreciation of medieval ruins and "Celtic temples." More tangibly, it helped to make images of the Middle Ages and their physical trappings a part of British culture. These images remained in Britons' minds long after the initial force of romanticism was spent. The first of Walter Scott's "Waverly Novels" was published in 1814, and in the next twenty-five years, 266 paintings based on his works alone were hung in London galleries; in 1820, the year after "Ivanhoe" appeared, London theatergoers were offered six different stage adaptations of it. After their 1828 opening, the armories at the Tower of London drew up to 40,000 visitors a year with their displays of medieval armor and weapons.[21] The popu-

20. The historical literature on the rise of the English middle class is enormous and complex. Here, I have drawn primarily on H. Perkin, *The Origins of English Society* (1969). On the natural history craze and "rational recreation," see D. E. Allen, *The Naturalist in Britain* (1976); L. Barber, *The Heyday of Natural History* (1980); and L. Merrill, *The Romance of Victorian Natural History* (1989).

21. Piggott, "Archaeological Societies," 183–84. Discussions of the medieval motif in Victorian culture include M. Girouard, *The Return to Camelot* (1981); A. Chandler, *A Dream of Order* (1970); and A. D. Culler, *The Victorian Mirror of History* (1985), chap. 7.

larity of medieval themes in art and literature helped to do for the secular remains of the past what the Oxford movement would do for the sacred ones. It transformed them, in the public's mind, from unappealing ruins to relics of a glorious, fascinating past.

The romantic movement also influenced archaeology by changing the study of history. Romantic historians in both Germany and England rejected the eighteenth-century philosophes' search for generalized patterns and universal laws of development in the human past. Instead, these turn-of-the-century scholars embraced a style of history that emphasized detailed studies of particular times and places and an awareness of the differences between them.[22] As practiced by popular writers such as Walter Scott, this style of history was founded "in the local and particular, in ballads and traditions, in abbeys and castles, in the antiquities and landscape of a region known deeply and personally."[23]

This affirmation of the importance of detailed local studies stimulated the "rediscovery of Britain" that interest in the Middle Ages had inaugurated.[24] Romantic historians argued that history could be done not only in Athens and Rome, London and Paris, but also in Penzance, Tillbury, and York. The same principle applied, by extension, to archaeological studies of the past. Archaeology—once virtually synonymous with long, expensive voyages to Egypt, Greece, and the Holy Land—now included day trips to ruined abbeys and evenings spent examining the local parish registers. By widening the study of the past, the romantic historians democratized it and thus brought it within reach of the increasingly receptive middle class.

The Oxford movement, which flourished from the late 1830s until the mid-1840s, was the creation of a group of young Oxford-trained theologians who hoped to reform the Church of England. In a series of pamphlets titled "Tracts for the Times," they emphasized the continuity of Anglicanism with Roman Catholicism, attempting to close a gap that liberal bishops had steadily widened since the turn of the century. These pamphlets also

22. H. Trevor-Roper, *The Romantic Movement and the Study of History* (1969).
23. Piggott, "Archaeological Societies," 185.
24. Hudson, *Social History of Archaeology,* chap. 2; the phrase is Hudson's.

called for a return to traditional forms of worship and particularly for a new emphasis on the importance of the "visible church." The movement saw traditional (that is, explicitly Catholic) forms of church architecture, ritual, costume, and liturgy as links not only with the pre-Reformation church, but also with the church of Saint Augustine.[25]

The Oxford movement was not universally successful; the popularity of Methodism and other dissenting sects blunted its effect in the north and midlands. In the south and west, however, Anglican clergymen—particularly those from rural parishes— flocked to the Oxonians' banner. The result was not just a new interest in church history, but an active campaign to renovate parish churches and restore them to their pre-Reformation glory. Virtually all of the restoration work that followed was carried out by parish members: remounting fallen sculptures, removing generations of whitewash and plaster from stonework, reclaiming fonts that had become planters and brass memorials that had become stove tops. The restoration work gave hundreds of parishioners an intimate acquaintance with the details of Renaissance and medieval church architecture and, through memorials and parish registers, with the history of their home towns.[26]

The romantic historians and the Oxford reformers encouraged the English middle class to rediscover their own heritage; the growth of Britain's railway network made that rediscovery possible. The greatest railway boom in British history took place between 1844 and 1847, when 2,000 miles of track were opened. When the boom ended, a quarter of a million men were working on 6,455 more miles. By 1852, the only sizable English towns not served by a railway were Hereford, Yeovil, and Weymouth.[27] It is not an accident that the railway boom coincided almost exactly with the flowering of the archaeological societies. The availability of fast, reliable transportation turned long, tiring excursions into easy overnight jaunts or day trips and made once-inaccessible sites into potential destinations for a holiday trip. As a result, a

25. R. K. Webb, *Modern England*, 2d ed. (1980), 233–36. For a more detailed discussion, see O. Chadwick, *The Victorian Church* (1967–70), 1: 212–21.

26. Piggott, "Archaeological Societies," 175–84.

27. P. Deane, *The First Industrial Revolution*, 2d ed. (1979), 172–73.

would-be archaeology enthusiast's horizons were limited only by
the extent of local branch lines and his ability to pay the fare.[28]

In 1810, the average archaeologist had been a wealthy gentleman living in London and collecting classical antiquities to display in his home. His counterpart of 1840 was a clergyman, physician, or shopkeeper living in the southern English countryside
and studying local ruins and documents in an attempt to learn
more about the history of his home town.[29] The county archaeological societies of the 1840s and 1850s were designed with this
new type of archaeologist in mind. Most had membership requirements much less stringent than those of the Society of Antiquaries, encouraging the participation of working-class and fe-
male enthusiasts. The county societies made it possible for their
members to gather regularly without making the long journey to
London and to meet others who were interested in the same
town or county. Taking advantage of the railways, the societies
organized joint meetings with other societies in neighboring
counties and field trips to points of archaeological interest. In
most cases, a county society's long-term goal was the compilation
of a county history, which would draw on legendary, documentary, and material evidence and range from the arrival of the earliest inhabitants to the present.[30]

The founders of the two new national societies, the British
Archaeological Association and the Archaeological Institute,
hoped to bind the local societies together into a national community of archaeologists. The first issue of the *Archaeological Journal,*
published before the institute's founders broke away from the
association in 1845, set down the principles that would eventually guide both the association and the institute. In it, Albert Way
took pains to point out that there would be no annual dues required of members, "the observation of such facts as may present
themselves, and the contribution of them toward the common
stock of knowledge being all that is expected."[31] The founders

28. Esther Moir's history of upper-class tourism in England cites 1840 as
the year in which genuinely "democratized travel" began. *The Discovery of Britain*
(1964), xiv–xvi.
29. For 1810 see Joan Evans, *Society of Antiquaries,* 215–24; for 1840 see Levine, *Amateur and Professional,* 48–51.
30. Levine, *Amateur and Professional,* 46–48.
31. [A. Way], "Introduction," *AJ* 1 (1844): 4.

hoped that their new organization would become the center of a network of correspondents in all parts of the country. The society's council, meeting fortnightly in London, would serve both as a clearinghouse for information sent in by the members and as a source of advice on proper technique and answers to difficult archaeological questions. Each year, in the style of the British Association for the Advancement of Science, the society would hold a general meeting in a different provincial town, bringing as many members as possible together for field excursions, general lectures, and sectional meetings devoted to specific interests.[32]

The founders of the new archaeological societies were motivated by more than altruism, however. They were equally, if not primarily, driven by their belief that the Society of Antiquaries had outlived its usefulness as an institution. Albert Way, secretary first of the British Archaeological Association and later of the Archaeological Institute, wrote in 1844 that "the Society of Antiquaries . . . although of a national and distinguished character, no longer fully supplies the exigencies of the occasion."[33] Thomas Wright, a founder and leading member of the association, was more blunt as he looked back a quarter-century later: "In England, at that time [1844], the long-established Society of Antiquaries was remarkable chiefly for its apathy to the interests of the science. In every part of the kingdom, interesting monuments were perishing from neglect or from actual violence, and no hand was held out to save them." To make matters worse, Wright argued, society members were likely to ludicrously misinterpret those objects that were saved. "Of those who did look upon such objects as possessed of some interest, few really appreciated them, or imagined that the study of antiquities had anything of science in it; but, each individual who obtained any single object of this description, seemed to think that by the mere possession of it he was qualified for giving an opinion and starting a theory upon it."[34]

Wright, Way, and the others intended the association and the

---

32. Ibid., 4–6; Levine, *Amateur and Professional*, 46–50; Joan Evans, "Archaeological Institute," 5–6.

33. [Way], "Introduction," 2.

34. T. Wright, "On The Progress and Present Condition of Archaeological Science," *JBAA* 22 (1866): 66–67.

institute to be the means of solving these problems—pulpits from which to spread the gospel of a new, scientific archaeology. The new archaeology, which both national societies promoted vigorously for the next half-century, blended the intellectual commitments of the local societies with a new attention to precision and scientific rigor. These methodological prescriptions apparently found a receptive audience, for the bulk of the papers in both the national and local archaeological journals reflect them. Ironically, the Society of Antiquaries also began to adopt the new approach after 1846—the year that reform-minded Phillip Stanhope became its president.

Historical Archaeology: Aims and Methods

The principles that historical archaeologists first espoused in the mid-1840s changed very little in the succeeding forty years. The establishment of human antiquity in 1858–59 displaced historical archaeology from the cutting edge of archaeological science but left the principles virtually unchanged. The prescriptive articles that appeared in the historical archaeologists' journals during the 1840s differ little from those that appeared in the early 1880s. Not only are the ideas themselves the same, but the metaphors and key words used to describe those ideas are strikingly similar. The same uniformity of style and content applies to the papers published in the three national journals: *Archaeologia*, the *Archaeological Journal*, and the *Journal of the British Archaeological Association*. By surveying all three publications over a period of several decades, it is possible to develop a clear picture of how historical archaeologists defined their science.[35]

Historical archaeology was founded on four basic principles. The first three were shared by both the local and the national societies: the importance of geographically narrow, temporally deep studies; the goal of holistically reconstructing the past; and the use of the widest possible variety of evidence. The last prin-

35. Levine discusses the social structure of historical archaeology in depth but its intellectual commitments only briefly (*Amateur and Professional*, 13–23, 70–74). Marsden provides a useful summary of historical archaeologists' names, careers, and major publications but little interpretation of their work (*Pioneers of Prehistory*, 25–35, 53–61).

ciple, a strict commitment to inductive methods, distinguished the reform-minded national societies from their local counterparts. The historical archaeologists' research program reflected all four principles, studying, in intimate detail, the entire history of regions that were seldom larger than an English county. Such an approach to the past echoed the romantic historians' emphasis on local studies and the interests of the county archaeological societies. The leaders of the national societies hoped to create a national history of Britain by linking local studies together like squares in a quilt, but plans for such a project remained vague. Thomas Wright's *The Celt, the Roman, and the Saxon*, first published in 1852, was the closest it came to fruition.

Historical archaeology's emphasis on the importance of local studies was more than a reflection of the interests of the county societies. The leaders of the historical archaeology movement insisted that local details were the key to understanding even national history. Artifacts should remain in local museums, Thomas Wright argued, because a local collection can illustrate both the general character of a vanished society and one of its many local variations. "Studied in one great centralized museum, where the local character of the objects is generally lost, we learn to appreciate what is Roman, or what is Saxon, or what belongs to any other particular period, but the real historical information we gain extends very little further."[36] The results of such a homogenizing study, Wright concluded, will be a picture of the past that is vaguely accurate in its outlines but wrong in most of its important details.

The historical archaeologists' preference for studies of change over time in a particular region also informed their view of archaeology's relationship with history. The relationship was frequently summed up in the statement that "archaeology is the handmaid of history,"[37] but it was more complex than the simple subservience that those words imply. For historical archaeolo-

36. Wright, "Progress," 69.
37. Variations on this theme include: E. Oldfield, "Introductory Address," *AJ* 10 (1852): 2–3; J. C. Bruce, "The Practical Advantages Accruing from the Study of Archaeology," *AJ* 14 (1857): 7; Lord Lytton, "Inaugural Address," *JBAA* 26 (1870): 22.

gists, archaeology and history were distinct but complementary ways of understanding the past—one based primarily on material artifacts, the other on documents. In theory, both disciplines covered the entire range of human history; in practice, one or the other took precedence, depending on the era being investigated and the types of evidence it offered.

Though willing to allow historians the leading role in writing the history of the last nine centuries, historical archaeologists believed that they, too, had an important contribution to make. They cast themselves in the crucial role of fact gatherer. Writing in 1852, Edmund Oldfield defined archaeology's purpose as "firstly, the discovery of evidence primary or collateral in proof of what is emphatically termed 'History'" and noted that archaeology, in fulfilling this function, "acts simply as purveyor to another, though kindred science."[38] It was among the archaeologist's self-assigned tasks to "prepare for historians the literature of documents generally," and to "ascertain the value of this unedited material in reference to what is already incorporated with printed literature, how far it suggests new views, supplies new facts, illustrates, corroborates, or disproves something previously acknowledged or disputed."[39]

The archaeologist held responsibility not merely for collecting facts from which historians could generalize, but for using those facts to independently test the historians' generalizations. George Tomline exulted, in 1865, over the "glee" that an archaeologist felt in "knocking a historian off his pedestal," and pulling down "theories which will not bear the scrutiny of our more intelligent and more abundant knowledge."[40] In 1860, another president of the British Archaeological Association cautioned that historians would have "avoided errors, and have arrived at juster conclusions, had they made fuller allowance for local and personal details."[41] Archaeologists, of course, saw these details as the heart of their science.

38. Oldfield, "Introductory Address," 2; [R. Westmacott], "Introduction," *AJ* 7 (1850): 5.

39. C. Newton, "On The Study of Archaeology," *AJ* 8 (1850): 10.

40. G. Tomline, "Inaugural Address," *JBAA* 21 (1865): 1; also see D. Croker, "On the Advantages of the Study of Archaeology," *JBAA* 5 (1849): 288.

41. Earl of Carnarvon, "The Archaeology of Berkshire," *JBAA* 15 (1860): 23.

In the Roman and Saxon periods of British history, where documents were scarce, historical archaeologists regarded their relationship with historians as an equal partnership. Extant documents, such as the narratives of Tacitus and Julius Caesar, were the province of the historians, but they could only provide a limited picture of the past.[42] The best way to overcome these limitations, the historical archaeologists maintained, was to rely on material remains: the Roman buildings, roads, inscriptions, coins, and household goods found throughout Britain. Archaeology clarified and extended the story told by the written evidence, fleshing out the bare skeleton of the documents.

The historical archaeologists' choice of subjects reflected their reliance on texts and their belief in a natural alliance between archaeology and history. Between 1850 and 1875, papers on medieval and later subjects made up 60 percent of *Archaeologia* and 70 percent of both the *Archaeological Journal* and the *Journal of the British Archaeological Association* (see App. 1). Papers on the Roman and Saxon eras—periods represented by large numbers of artifacts and limited numbers of documents—were comparatively scarce. Papers on the pre-Roman (or Celtic) era, for which artifacts were the only records, were the least common of all. Historical archaeologists were, with few exceptions, drawn to periods of Britain's past that offered extensive documentary records and opportunities for detailed reconstructions.

The goal of creating such reconstructions—the second major element of the historical archaeologists' research program—was rooted not in history but in natural science. In the early 1830s, a group of English geologists led by William Buckland added a new dimension to the study of Earth history by pioneering the reconstruction of ancient environments. In the layers of rock that made up the Earth's crust, Buckland and his colleagues saw evidence of a succession of "ancient worlds," each with characteristic plants, animals, and climatic conditions. The geologist's goal was to use both fossils and the rocks that contained them to reconstruct these ancient worlds in detail—to do for vanished environments what Georges Cuvier had done for extinct animals. Their recon-

42. Bruce, "Practical Advantages," 7.

structions, both textual and visual, captured the British public's imagination.[43]

The historical archaeologists, too, drew their inspiration from Cuvier's work. They sought to expand lists of names and dates into tableaux filled with three-dimensional human beings, as geologists and paleontologists transformed dry lists of rock layers and fossil species into panoramas of ancient landscapes. Edmund Oldfield, who saw archaeology's first job as the gathering of facts for historians, nevertheless recognized the importance of such reconstructions. He argued that they illustrated "personal life among our ancestors, in points of which national History takes no account, as lying, in a manner, off its highway. Archaeology here no longer holds a merely ancillary position, but itself rises to the level of History, as it furnishes the only memorial of what the great masses of mankind individually were, and did, and thought, and felt, in former ages."[44] If history would provide the framework, then archaeology would undertake the reconstruction.

The historical archaeologists drew both implicit and explicit analogies between their reconstructions of the past and Cuvier's recreations of extinct life. Without archaeology's aid, one popular metaphor suggested, "history would as little represent the particular time it endeavoured to recall as the drawing of a skeleton would represent the features and form by which the individual human being was recognized while in life. It is to the skeleton of a former age that archaeology restores the flesh and the sinews and the lineaments that distinguish it from the countless centuries of which it is a link."[45] C. W. Hoskyns, claiming that "it is in exact analogy with such a recreative power as [Cuvier's] that archaeology develops its most useful results," took the analogy even further. For him, archaeology "restores to the skeleton of history" not only its flesh and sinews but even "the very bearing and expression, which each age wore as it passed by."[46]

43. Rupke, *The Great Chain of History* (1983), 130–49.
44. Oldfield, "Introductory Address," 2–3.
45. Lytton, "Inaugural Address," 22.
46. C. W. Hoskyns, "Inaugural Address," *JBAA* 27 (1871): 1–45, on p. 24. Also see Marquis of Northampton, "Inaugural Address," *AJ* 2 (1846): 302–3; and T. J. Pettigrew, "On the Study of Archaeology," *JBAA* 6 (1850): 163–64.

Historical archaeology was capable of such feats, its practitioners believed, because accurate reconstruction hinged on an
extensive knowledge of the smallest details of a particular age.
Those details, ignored or glossed over in the documents used by
traditional historians, were to be found instead in the material
objects that archaeologists studied. So, the Earl of Carnarvon told
his audience in 1859, "nothing which can aid or illustrate [histo-
ry's] teaching is below her dignity. And thus the pattern of a
chair, the fashion of a tapestry, the roll of a parchment, the turn
of a letter, which are the specialities in the craft of the archaeologist, become invaluable to the historian."[47]

Buckland and his circle, who sought to reconstruct the ancient land- and seascapes of the geological past in as much detail
as possible, cast their nets wide when they gathered evidence. In
addition to animal and plant fossils, they examined gnawed
bones, coprolites (fossilized excrement), and the character of the
sediment in which all were imbedded.[48] The historical archaeologists, studying a more recent period and faced with a much
greater variety of evidence, cast their nets even wider. Charles
Newton, addressing the Archaeological Institute in the summer
of 1850, claimed that the archaeologist's true motto should be
"*Homo sum, humani nihil a me alienum futo* [I am a man, nothing
human is alien to me]."[49] A catholic mixture of sources was one
of the hallmarks of historical archaeology, and a good archaeologist was expected to collect, describe, and assess the significance
of any artifacts that came before him.

Ambitious local societies sought to enlist the general public
as well as their own members in the search for the widest possible
range of artifacts. They advertised their interest in, and hinted at
their willingness to pay for, "relics bearing the stamp of antiquity,
of whatever kind." A circular printed by the Moray Literary and
Scientific Association in 1851 exhorted members to inform "sur-
veyors, contractors, and others engaged in improving or draining

47. Carnarvon, "Archaeology of Berkshire," 24. Also see [Westmacott], "Introduction," 4; and archbishop of York, "Inaugural Address," *AJ* 24 (1867): 88.
48. Rupke, *Great Chain of History*, 133–48.
49. Newton, "Study of Archaeology," 24. Also see [J. M. Kemble], "Introduction," *AJ* 6 (1849): 2; and A. Conyngham, "Presidential Address," *JBAA* 5
(1849): 285–88.

Land" that any antiquities they unearthed would be "held in due estimation" by the association, and that "this will be the best market to which any poor man can bring such articles of curiosity as he may find in the district." The circular concluded with a list of antiquities such as "warlike instruments," sculpture, household implements, "sepulchral offerings," and manuscripts.[50] At a time when most sciences were growing increasingly specialized, generalism was a way of life in the Moray Association and in British archaeology as a whole.

Historical archaeologists believed that material objects were only the beginning of a proper reconstruction of the past. Lord Lytton listed "guard[ing] from oblivion the myths, the traditions, the legends of former days" among their duties, and "the preservation of many a pure and sacred well-spring of poetry and romance" among mid-Victorian archaeology's accomplishments.[51] Archaic words, inflections, and idioms—vestiges of earlier languages or dialects that survived in rural speech—also formed an important part of the historical archaeologists' science.

Because most of them studied literate societies and because they saw archaeology as a close ally of history, the historical archaeologists also made extensive use of written documents. Some members of the community, such as Charles Newton, saw it as the archaeologist's job to prepare newly discovered documents for historians by comparing them with those already known. Others, particularly those interested in churches and municipal buildings, used official records and drawings of such structures as an aid in understanding them.[52] A third group apparently regarded old documents as an interesting and important part of the past in their own right; at least five or six were printed in each year's edition of the *Archaeological Journal* and the *Journal of the British Archaeological Association.*

Nearly all Victorian scientists claimed to follow the empiricist methods outlined by Francis Bacon. None, however, sang empiricism's praises more loudly or more consistently than the histori-

---

50. Patrick Duff, Moray Literary and Scientific Association Circular, 11 March 1851, Falconer Papers, DAV/ZF492.

51. Lytton, "Inaugural Address," 23.

52. York, "Inaugural Address," 85.

cal archaeologists. Though they invoked neither Bacon's name nor the names of more recent philosophers, the formal speeches at their annual meetings, as well as both the rhetoric and content of their "everyday" papers, reflect a near-obsession with inductive, empiricist science.[53]

Facts, the historical archaeologists proclaimed again and again, were their stock-in-trade. Archaeology, they declared, "takes nothing for granted, owes little to theory or hypothesis; but points to early vestiges, substantive records, visible and tangible existences."[54] Presidents of the national archaeological societies repeatedly emphasized the value of empiricism in their addresses at annual meetings. The proper business of the archaeologist was the collection, comparison, and organization of facts; once enough facts had been gathered, they would somehow fuse into an accurate, impressively detailed picture of the past. For Lord Northcote, this process had an almost mystical quality to it. "You may find any number of small points, each insignificant, apparently absurd, if you take it by itself, yet if you put them together, compare them, collate them with what has been discovered in other parts of the country, they produce, by degrees, first doubt, then suspicion, and then a moral certainty which almost amounts to the strength of a demonstration."[55]

The historical archaeologists' delight in facts was matched by their suspicion of—or outright hostility to—theories. The most extreme form of this position held that archaeology should be completely atheoretical. Like early Victorian students of natural history, they believed that the essence of what they studied lay not in universal laws but in particular details.[56] "We ask for facts, and evidence of facts," Lord Houghton told a meeting of the BAA in 1864, "and we are content to leave to others the responsibility of their own conclusions. Most people of any ingenuity can dis-

53. Several aspects of the historical archaeologists' method resemble the "agonistic style" of early Victorian geologists such as George Greenough and Henry de la Beche. See M. J. S. Rudwick, "Cognitive Styles in Geology," in *Essays in the Sociology of Perception*, ed. M. Douglas (1982), 227–29.

54. Hoskyns, "Inaugural Address," 25.

55. S. H. Northcote, "Inaugural Address," *JBAA* 18 (1862): 10. Also see [Kemble], "Introduction," 2.

56. Merrill, *Victorian Natural History*, 11–13.

cover in history whatever they wish to find there; and if you choose to make a bad use of the materials with which we supply you, do it at your own risk, and do not lay the blame on archaeology."[57] The Earl of Mount Edgcumbe, quoting archaeologist William Borlase in 1877, argued that archaeology is "of the greatest value when it seeks to rest the vapoury superstructure of theory or tradition upon the firm basis of observed fact."[58]

The majority of historical archaeologists, while more receptive to theories than were Houghton and Mount Edgcumbe, agreed that theories be put in their proper place. The archbishop of York spoke for this majority when he outlined the relationship between fact gathering and theory building in 1866. "To avoid a groundless theory seems to have become, as it were, part of the moral code of the archaeologist. The time for theories, it seems to be admitted, begins when the collection of facts has been as large and general, and as exhaustive as the subject seems to admit." A science has attained perfection when "facts have been abundantly supplied and theory has been used with soberness and yet with bold sagacity for their explanation."[59] Describing the founding of the British Archaeological Association, Thomas Wright emphasized that historical archaeology's disdain for fanciful interpretations distinguished it from earlier, unscientific archaeology. He told an 1866 audience that "it was some time before we could sufficiently impress upon our correspondents that the collecting of facts must precede the deduction of science from them."[60]

The papers published in the three national archaeological journals demonstrate that rank-and-file historical archaeologists shared their leaders' commitment to empiricism. Virtually every paper published between 1850 and 1875 involved descriptions of new evidence, comparisons of new and existing evidence, or

57. Lord Houghton, "Inaugural Address," *JBAA* 20 (1864): 4.

58. W. C. Borlase, *Naenia Cornubiae* (1872), quoted in Earl of Mt. Edgcumbe, "Inaugural Address," *JBAA* 33 (1877): 4.

59. York, "Inaugural Address," 85.

60. Wright, "Progress" (no. 34 above), 67. A number of speeches given in the mid-1840s contrast historical archaeology's new, "scientific" approach with the work of earlier antiquarians, who used theory and speculation too liberally. Two examples are [Kemble], "Introduction," 1–2; and [Westmacott], "Introduction," 3.

both. The descriptions met high standards: they were detailed, precise, and well illustrated with engravings, maps, and diagrams. Only rarely, however, were they combined with any attempt at explanation or generalization.[61] When the archbishop of York told his 1866 audience that "in the transactions of . . . our English Societies there is a remarkable caution and sobriety," he understated the case.[62] Victorian scientists often sought to validate their work with claims that they had "collected facts on a wholesale scale."[63] The historical archaeologists actually did so, year after year, for decades.

Reverend William Greenwell's 1865 paper on the "ancient grave-hills" of north Yorkshire epitomizes the historical archaeologists' published work.[64] Greenwell was conversant with all aspects of British archaeology but specialized in the study of the barrows, or ancient burial mounds, found throughout the British countryside. His 1865 contribution to the *Archaeological Journal* was only a milestone on the road to his 750-page work, *British Barrows* (1887).[65] Greenwell's paper was extremely long by the standards of the journal that published it: forty-four pages when most papers ran less than ten. Of these forty-four pages, all but the first two were solid masses of facts. For each of more than a dozen sites, Greenwell described the barrow's shape and structure in detail, then turned his attention to the bones and artifacts it contained.

The investigation of British barrows was already a well-established part of archaeology when Greenwell wrote his paper. His argument that yet another paper on the subject would offer more than a repetition of well-known facts thus highlights the value that he placed on fact gathering. No two barrows, he began, "present quite the same features, and each one that is examined is valuable, either as a confirmation of some view not yet based

61. This characterization is based on a comprehensive survey of papers on pre-Roman subjects published between 1850 and 1875, along with a more random survey of papers on later subjects.

62. York, "Inaugural Address," 86.

63. The specific wording used here is from Charles Darwin's *Autobiography*, ed. N. Barlow (1958), 119.

64. W. Greenwell, "Ancient Grave-Hills in the North Riding of Yorkshire," *AJ* 22 (1865): 97–117, 241–65.

65. Discussed in Marsden, *Pioneers of Prehistory*, 60–62.

on a sufficiently exact or wide foundation, or as giving some new fact that may modify, or perhaps destroy the theory which, in such matters, we are sometimes obliged to erect."[66]

Greenwell's two-page introduction neatly summarized the basic principles of the historical archaeologist's method. Facts are the archaeologist's first concern, and knowledge of the past can only grow from vast collections of them. Even in the study of barrows, where data had (by 1865) been collected for nearly a century, each new fact is valuable, and necessary. The particular must always take precedence over the general, and the truth resides in local details rather than regional or national syntheses. Theories are acceptable but always suspect and susceptible to modification or destruction by a new fact; when they are introduced, it is only because the archaeologist is obliged (presumably against his better judgment) to create them. In any event, theories make only a minor contribution to the search for truth, which is a tightly woven "fabric" rising from a "firm substructure" of "well-sifted and oft-recurring detail."[67]

Discourses on method like the one in Greenwell's paper were more than rhetorical devices designed to invoke the spirit of Francis Bacon; they were statements of deeply held beliefs. The empirical method they described was the intellectual thread that linked the components of historical archaeology. Faith in such methods served as the organizing principle behind the British Archaeological Association and the Archaeological Institute, which were designed to serve as collection centers and clearinghouses for new archaeological data. Definitions of archaeology's relationship to history reflected the same faith, as did the ways in which archaeologists presented their findings. Finally, faith in empiricism reinforced the archaeologists' belief that the particular must always take precedence over the general and justified their unwillingness to undertake a national or even a regional synthesis. The historical archaeologists believed that their commitment to empiricism, more than any other factor, set them apart from their predecessors and from the community of "prehistoric archaeologists" that emerged in the 1860s.

66. Greenwell, "Ancient Grave-Hills," 97.
67. Greenwell, "Ancient Grave-Hills," 97–98.

Historical Archaeology and Prehistory

When Daniel Wilson coined the word *prehistory* around 1850,[68] the idea of an age before recorded history was well established in British archaeological thought. It was common knowledge among educated Britons that the Celts had lived in Britain long before Julius Caesar's invasion of 55 B.C., and artifacts from the once-flourishing Celtic civilization had been found throughout the British Isles. The Celts' polished stone weapons (recognized as such since the late seventeenth century)[69] turned up regularly in building excavations; Celtic burial mounds dotted the countryside; and Stonehenge, attributed to the Celtic priesthood known as the Druids, loomed over Salisbury Plain. Prehistory was a briefer, more shadowy period for the archaeologist of 1850 than for his counterpart of 1890, but it was nonetheless a legitimate part of archaeology.

Despite this, the historical archaeologists took little notice of prehistoric sites and artifacts. Between 1850 and 1859, the three national archaeological journals published 629 papers, of which only 34 (5.4 percent) dealt with prehistoric subjects.[70] In contrast, the same ten-year period saw the publication of 449 papers (71.4 percent of the total) on subjects from medieval and later times. The historical archaeologists' apathy toward the accumulating evidence of human antiquity reflected their lack of interest in prehistory—a lack of interest conditioned by the methods and goals they shared.

Compared to that for the Middle Ages, the archaeological record for British prehistory was depressingly fragmentary. In the 1850s, it amounted to no more than a few hundred artifacts, unearthed largely by chance at sites scattered all over the British Isles. Worse, those artifacts represented only a small, skewed sample of the prehistoric Briton's material culture. The ravages of time, dampness, and acidic soil wiped away all traces of wood, leather, and textiles and often reduced iron to a black smear in

68. Daniel, *150 Years* (no. 3 above), 83.
69. D. K. Grayson, *The Establishment of Human Antiquity* (1983), 5–7.
70. Cf. M. Bowden, *Pitt Rivers* (1991), 150–51; the two societies' journals offer little support for Bowden's claim that the institute was more interested in prehistory than was the association.

the ground; pottery, if it survived the elements, was frequently smashed to fragments by the plow or ditchdigger's shovel that accidentally uncovered it. The tangible remains of prehistory consisted, in the 1850s, primarily of stone and bronze implements, some pottery, a few gold ornaments, and those stone monuments that had not yet been dismantled in order to clear fields or provide building material.

For historical archaeologists, who made it their goal to build up a mosaiclike picture of the past using facts from a wide variety of sources, this paucity of evidence posed a nearly insurmountable problem. The entire stock of facts about prehistoric society in Britain was small, and the collections from particular localities were smaller still. By the standards of historical archaeology, such small collections of facts were scarcely better than no facts at all— they constituted, not the basis of a mosaic, but a jigsaw puzzle with 97 percent of its pieces missing. Many of the artifacts that had survived provided no clues to their function and bore no resemblance to anything with which modern society was familiar. Discussing Celtic artifacts, Thomas Wright wrote that "the solitary dagger, with the few fragments of pottery and two or three beads or pins, can give us no satisfactory notion of the dress or riches of the person who wore them."[71] The Middle Ages, by contrast, offered readily identifiable artifacts by the thousand, along with written documents, art, legends, songs, and religious relics. Unlike prehistory, the Middle Ages held out the promise of an accurate archaeological and historical reconstruction.

The absence of an internal framework that could be used to give order to scattered facts further diminished prehistory's appeal to historical archaeologists. Artifacts from medieval or later periods could be dated confidently as long as they included some inscription or some quirk of design that could be referred to a particular era or reign. Even without such identifying marks, archaeologists could date them based on references in written documents, or on the artifact's association with other objects (frequently coins) of known date. Prehistory, by definition, lay beyond the reach of the intricately divided and subdivided time

71. T. Wright, *The Celt, the Roman, and the Saxon* (1852), 80. The statement was carried over, unaltered, into 2d (1861) and 3d (1875) editions.

scale defined by written documents and had no equivalent time scale of its own. As a result, the absolute dates of prehistoric artifacts could not be determined, and theories about their relative ages could not be verified. Thomas Wright, discussing primitive stone tools in the 1850s, was at a loss to settle their ages. It might range, he said, "from a limit we have no means of fixing" to Anglo-Saxon or even later times.[72]

The Three-Age System, developed in Scandinavia, provided a method of estimating the relative ages of prehistoric artifacts, but English archaeologists were hesitant to adopt it. The system was created by Christian Thomsen of Copenhagen's Royal Museum of Northern Antiquities in 1819 and popularized by Sven Nilsson and J. J. A. Worsaae. Thomsen published the first clear statement of it in an 1836 museum guidebook, and Lord Ellesmere produced an English translation in 1848.[73] The Three-Age System gave a measure of internal structure of prehistory by dividing it into Stone, Bronze, and Iron ages, each named for the material most frequently used during the period for making tools and weapons. The three ages occurred in the same sequence in the prehistory of every European country, Thomsen argued, though the duration of each age—and of the transitions between them—would vary from area to area. An archaeologist who embraced the system could thus conclude that a bronze implement from a given area was likely to be more modern than a stone implement from the same area.

Archaeologists in Scandinavia and other parts of Europe accepted the system readily, and a few British archaeologists regarded it with cautious optimism.[74] Most members of the British archaeological community had serious reservations about Thomsen's ideas, however. True to their Baconian creed, they regarded the Three-Age System as an unwarranted generalization that the

72. T. Wright, "On the Remains of a Primitive People," in *Essays on Archaeological Subjects* (1860), 1: 6–7.

73. Daniel, *150 Years*, 38–42; also see B. Gräslund, "Thomsen's Three Age System"; and J. Rodden, "Development of the Three Age System," both in *Towards a History of Archaeology*, ed. G. Daniel (1981).

74. Daniel, *150 Years*, 40–48, 77–79. Edmund Oldfield was among the few historical archaeologists who saw the system as a potentially useful scheme; see Oldfield, "Introductory Address," 5.

facts did not support. T. J. Pettigrew, speaking to the British Archaeological Association in 1855, argued that "it will be seen that difficulties in regard [to the three ages] must necessarily arise from the instances occurring during their several periods of transition." J. M. Kemble offered his listeners at an 1857 meeting of the Royal Irish Academy a long list of cases where bronze implements had been used alongside iron ones, and stone alongside both. Many stone implements, he agreed, actually belong to the "earliest period of human culture," but many more do not. The fact that one implement is stone and another is bronze, Kemble concluded, says nothing about their relative ages.[75]

A common theme ran through Pettigrew's and Kemble's critiques of the Three-Age System and continued in the work of Thomas Wright.[76] Kemble and Wright presented lists of specific cases where tools of different materials were found alongside each other, and Pettigrew referred to the problem posed by transition periods. None of the authors criticized the system's internal logic. All three clearly believed that a simple enumeration of counterexamples represented a damning criticism of it—a belief completely in keeping with their commitment to empiricist methods. They argued that the archaeologist should "never be astonished to find a pet notion rudely dashed to the ground by a stroke of the pickaxe, or a turn of the shovel. Never should he be ready to sacrifice a fact, merely because it is hard to explain, upon the altar of a much more indefensible theory."[77]

Much of historical archaeology's popular appeal lay in its concern with reconstructing the history of particular regions. The local societies, focusing on the histories of their home counties, were driven by their members' local pride and desire to seek out and preserve whatever made their part of Britain unique. This appeal depended, however, on the existence of monuments, artifacts, or events that could be claimed by a particular county or town and serve as a focus for archaeological interest. The medi-

75. T. J. Pettigrew, "On the Antiquities of the Isle of Wight," *JBAA* 11 (1855): 184; J. M. Kemble, "On the Utility of Antiquarian Collections," *Proceedings of the Royal Irish Academy* 6 (1857): 462–80.

76. Wright, *Celt, Roman, and Saxon*, chap. 1, *Essays on Archaeological Subjects*, and "Progress."

77. W. C. Borlase, *Naenia Cornubiae* (1872), 6.

eval and Roman periods offered many such foci. Citizens of Glastonbury could take pride in their ruined abbey, those of Bath in their Roman baths, and those of Brixham in the site where William of Orange—soon to be King William III—arrived from Holland in 1688.

Prehistory, by contrast, offered almost no such sites. Warwick Castle and Canterbury Cathedral had clear geographical associations and well-defined places in history, but what connection could Celtic burial mounds on a Yorkshire moor have with the Yorkshire of 1850? Mysterious Celtic ruins might inspire a sense of awe if seen through romantic eyes, but they were unlikely to evoke a sense of local or national pride.[78] Lord Northcote, speaking to the British Archaeological Association in 1861, summed up an important distinction between the two kinds of archaeological relics. There are some, he said, "which are so old, which belong to a time so far bygone, that they excite little else than wonder. You find others which carry us continuously up to the present day, and seem to have a more living and present interest for us."[79] Historical archaeologists preferred, almost unanimously, to study monuments and artifacts from the latter category.

It is precisely this preference—for the wealth of evidence, detailed historical framework, and clear ties to the present offered by later periods—that drew historical archaeologists' attention away from prehistory. Those who did study prehistoric artifacts did so less as a specialty than as a sideline, and none published more than two papers on prehistory between 1850 and 1859. The thirty-four papers that appeared in the 1850s were not the product of an active attempt to recreate the earliest history of Britain. Rather, they were intended to describe new sites and artifacts as they came to light. The handful of archaeologists who wrote them believed that even a sketchy picture of prehistory still lay far in the future and were content, for the present, to gather the facts from which that picture would emerge.

78. J. Michell, *Megalithomania* (1982), discusses and illustrates the range of artistic and scientific reactions to such ruins.
79. Northcote, "Inaugural Address" (n. 55 above), 5.

THREE

Geologists and Human Antiquity to 1858

LOOKING BACK, at the end of the nineteenth century, British geologists saw the years between 1820 and 1860 as a "golden age" for their science. Such late Victorian geologists as Archibald Geikie and William Boyd Dawkins glorified their predecessors in print, praising them as intellectual giants and marveling at the range of questions they grappled with. The early Victorian leaders of the geological community were, their successors insisted, a breed apart. Geologists with the stature of William Buckland or Charles Lyell, Adam Sedgwick or Roderick Murchison would not come again. Geology had, Dawkins lamented, become the domain of specialists whose narrowly focused research could not illuminate the fundamental problems that had once been geologists' stock-in-trade.

The late Victorian image of Sedgwick, Murchison, and Lyell as larger-than-life heroes was a powerful rhetorical device, but there was more to it than rhetoric. The four decades between 1820 and 1860 saw the elucidation of a fundamental order for the world's rocks, the reconstruction of the dinosaurs and the world in which they lived, and the triumph of Charles Lyell's geological philosophy of actualism. In the same period, the first paid positions for geologists emerged in Britain: professorships at Oxford, Cambridge, and King's College; and posts with the Geological Survey. Buckland, Lyell, Sedgwick, Murchison, and others of similar stature were at the height of their intellectual powers between 1820 and 1860. The younger men who supplanted them as leaders of the geological community began their careers in the 1850s. Throughout this forty-year period, geology enjoyed an enormous, talented amateur following and unrivaled popularity among the members of the educated public. The intellectual and social structure of British geology during the final decade of the golden age set the terms on which the human antiquity problem was discussed and, in time, resolved.

Geologic Thought in the "Golden Age"

In the 1850s, British geology flourished, guided by a well-established paradigm. The paradigm—built around fieldwork, empirical methods, and studies of stratigraphy and paleontology—emerged between 1800 and 1820. Its introduction, along with the founding of the Geological Society of London and the establishment of geology professorships at Oxford and Cambridge, laid the foundations of Victorian geology.[1] Because neither the paradigm nor geology's claims to disciplinary status were in doubt by the 1850s, few explicit statements of goals and methods appear in the British geological literature of that period. Such statements are implicit throughout the literature, however—notably in textbooks and other works intended for the general public.

Programmatically, geology included the study of the Earth's structure, composition, and surface features as well as the history of its rocks and of the fossils they contain. In practice at its heyday, however, British geology was virtually synonymous with stratigraphy and paleontology. Most Victorian geologists spent their careers mapping the extent and describing the contents of the rock layers, or strata, that make up most of the Earth's land surface. Eminent geologists such as Lyell or Murchison grappled with broad, theoretical questions—the rate of geologic change, and progression in the fossil record, for example—but only their encyclopedic knowledge of stratigraphy gave them license to do so. For dozens of lesser geologists, stratigraphy was the alpha and the omega of a geological career. Many of them devoted a lifetime to understanding the strata laid down in a particular time period or region, as Hugh Miller did with the Old Red Sandstone.

The everyday papers written by British geologists during the 1850s clearly reflected this concern with stratigraphy and paleontology. Seventy-two percent of those read at meetings of the Geo-

1. R. Porter, *The Making of Geology* (1980), chaps. 6 and 7.

logical Society of London and published in its *Quarterly Journal* were concerned with stratigraphy, paleontology, or a blend of both. By contrast, 10 percent of the papers in the *Quarterly Journal* concerned physical geology—landforms, or geological structures such as mountain belts and fault lines—and only 7 percent treated the chemical composition of rocks. The geological papers read to the British Association for the Advancement of Science also dealt primarily with strata and fossils. Sixty-one percent concerned stratigraphic and paleontological subjects, while physical and chemical geology accounted for 19 percent and 9 percent, respectively.

Mid-nineteenth-century popularizations, manuals, and textbooks of geology—many of them written by leading geologists—showed a similar preoccupation with stratigraphy and paleontology at the expense of other topics. John Phillips opened his 1855 *Manual of Geology* with a revealing definition of geology. He described geology as the science concerned with investigating "the ancient natural history of the earth," using "natural history" to denote a descriptive, taxonomic study distinct from "natural philosophy's" study of causes.[2] The first goal of geology, he continued, is to "determine by observation what phenomena of living beings or inorganic matter were formerly occasioned on or within the globe, in what order and under what conditions."[3] The remainder of the book carried out this plan in extraordinary detail; 528 of its 634 pages were devoted to strata and their fossils. The remaining hundred pages were divided between a discussion of changes in the Earth's surface and a discourse on the current state of geologic theory. Phillips offered his readers no systematic discussion of physical geology and no discussion at all of the chemical composition of rocks or minerals.

The other major popular works on geology that appeared in the 1850s—the fifth edition of Charles Lyell's *Elements of Geology*, the seventh edition of Gideon Mantell's *Wonders of Geology*, and the second edition of William Buckland's *Geology and Mineralogy*—took similar approaches. Mantell's work consisted primarily

2. Phillips' distinction is similar to Rachel Laudan's division of geology into "causal" and "historical" branches. See R. Laudan, *From Mineralogy to Geology* (1987), 1–16.

3. J. Phillips, *Manual of Geology* (1855), 2.

of detailed descriptions of strata and fossils that, like Phillips's, begin with the oldest rocks and work systematically toward the youngest. Lyell's work, aimed at students and less theoretical than his famous *Principles of Geology*, took the same approach. Buckland's two-volume book concentrated on paleontology despite its title, and, again, begins with the oldest fossils and concludes with the most recent ones. All these works were revisions of earlier editions from the 1830s and 1840s, but none represented a significant departure in structure or emphasis. The books' overwhelming concern with stratigraphy and paleontology was part of a decades-old tradition in British geology.[4]

The conception of geology as a science devoted almost exclusively to paleontology and stratigraphy extended beyond the boundaries of the geological community. An anonymous reviewer of the second edition of Buckland's *Geology and Mineralogy* began by noting that the book's lack of attention to mineralogy was not a serious problem because "mineralogy itself has in reality little to do with geology, so little that a man might be a first rate geologist—not a perfect one, certainly, but a very eminent and sound one—without knowing anything whatever of the science of mineralogy."[5] Another reviewer, discussing a new textbook written in 1858 by Joseph Beete Jukes of the Geological Survey, praised it for taking the unprecedented step of giving chemical and physical principles approximately equal space with paleontology and stratigraphy. Such a book, the reviewer claimed, would help to speed geologists' slow "awakening to the value of chemistry in determining the constitution of rocks," and to cure their "almost entire disregard of physics."[6]

Except in the hands of Lyell and a few others, British stratigraphy was virtually atheoretical. Geologists preferred to establish the relative order of strata before turning to questions about how quickly, and under what conditions, they had been deposited.[7] A similar approach dominated British paleontology. Separate at-

4. Mineralogy played a far larger role in geology on the Continent than it did in Britain. Laudan's *From Mineralogy to Geology* explores that role in detail.

5. Review of *Geology and Mineralogy*, *Dublin University Magazine* 54 (1859): 455.

6. Review of *A Student's Manual of Geology*, *Athenaeum*, 13 (February 1858): 203.

7. J. A. Secord, *Controversy in Victorian Geology* (1986), 4.

tempts by William Buckland and Henry De la Beche to make the reconstruction of ancient environments an important part of geology had, by 1850, collapsed without making a lasting impression on the geological community.[8] British paleontologists saw their principal goal as the reconstruction of the history of life on Earth—the establishment of when, in what order, and in what geographic areas new species had appeared. In the 1850s, as in earlier decades, most geologists and paleontologists saw strata and fossils not as vestiges of former worlds but as pieces in a three-dimensional physical puzzle.

Victorian geologists saw stratigraphy and paleontology as intimately linked because solving such three-dimensional puzzles required an intimate knowledge of both strata and fossils. Starting work in an unexplored area, the geologist would make a general survey, then examine each stratum in turn, noting its thickness, composition, fossil content, and physical relationship with the other strata. This done for each layer, he would use a mixture of these criteria to establish the order in which the strata had been deposited.[9] Where the strata were extensive, horizontal, and undistorted, establishing the succession was usually straightforward; geologists had recognized since the 1660s that the uppermost beds in an undisturbed sequence of strata were the youngest and the lowermost were the oldest. Undisturbed strata were rare, however. The strata were often folded, overturned, partially eroded, or covered by soil and vegetation, and geologists could establish their order of deposition only by noting where the fossils in each layer fitted into the succession of life on Earth.

The geologist's next goal, after unraveling a new area's stratigraphy and paleontology, was to correlate its strata with those of surrounding areas. Over short distances, it was theoretically possible to make such correlations on the basis of rock type alone, but the majority of geologists did not use this method regularly.[10]

8. N. A. Rupke, *The Great Chain of History* (1983), 267–72; J. A. Secord, "The Geological Survey of Great Britain as a Research School," *History of Science* 14 (1986): 223–75.

9. Secord, *Controversy,* 25–27.

10. The most famous exponent of this method was Adam Sedgwick. Secord, *Controversy,* 57–68; M. J. S. Rudwick, "Levels of Disagreement in the Sedgwick-Murchison Controversy," *QJGSL* 132 (1976): 373–75.

Over longer distances, the method broke down entirely. Strata laid down simultaneously in different parts of the country were often products of different environments and so were composed of entirely different types of rock. Most British geologists regarded environmental variations in fossils as less problematic, and so used fossils as time markers for both short- and long-distance correlations. This idea, first developed in the 1790s, firmly linked stratigraphy and paleontology in British geological thought.

The idea of "index fossils," developed simultaneously by William Smith in England and by Georges Cuvier and Alexander Brogniart in France, was based on the premise that all species had a finite lifetime.[11] Ideally, if a now-extinct species had a short "range"—if it had existed for only a brief time in the past—then any strata in which its fossils were found must have been deposited within that brief time. Practically, the incompleteness of the fossil record meant that correlations based on groups of fossil species were more reliable. The most useful groups were those in which the member species' ranges overlapped for only a comparatively brief time. Any stratum that contained all the member species must thus have been deposited during the period when the member species coexisted. The use of index fossils made it possible to correlate strata over enormous distances—across national and, in theory, even continental boundaries. A layer of shale in Oxford and a layer of sandstone in Yorkshire might have no obvious relationship to each other, but if they contained the same set of index fossils, then they must have been laid down in the same period. Further, if the Oxford shale lay at the bottom of one well-described sequence of strata and the Yorkshire sandstone lay at the top of another, the two sequences could be readily fitted together into a single, continuous sequence.

Beginning in the 1820s, British geologists set out to describe and classify groups of strata which, when linked together, would represent the entire span of geologic time. They divided the geologic past into five great eras, subdivided each era into periods, and split each period into epochs (fig. 3). Geologists attempted to fit first Europe's and then the world's strata into this framework

11. W. B. N. Berry, *The Growth of a Prehistoric Time Scale*, 2d ed. (1987), 53–59.

of eras, periods, and epochs. When complete, this minutely classi-
fied picture of the Earth's strata was intended to serve as an international standard. Even when working in an unexplored area of
a foreign country, a geologist could compare the fossils in a newly
described sequence of strata to the index fossils in the already-
documented sequence and know which subdivision of the geologic past the strata had been deposited in. The geologist could be
sure that, no matter how thick it was or what it was made of, a
stratum containing the clawed, hard-shelled invertebrates known
as eurypterids belonged to the lower Silurian period of the Paleo-
zoic era. A nearby group of strata containing the bones of small,

ERAS	PERIODS	Epochs
CENOZOIC	QUATERNARY	Recent
		Post-Pliocene
		Pliocene
	TERTIARY	Miocene
		Eocene
MESOZOIC	CRETACEOUS	
	JURASSIC	
	TRIASSIC	
PALEOZOIC	PERMIAN	
	CARBONIFEROUS	
	DEVONIAN	
	SILURIAN	
	CAMBRIAN	

Figure 3. The major subdivions of geologic time as British geologists
saw them in 1858 (based on Lyell, *Antiquity of Man*, 1863).

primitive dinosaurs would just as surely have been laid down later, in the Triassic period of the Mesozoic era.

British geologists saw fieldwork as the key to understanding the fundamental order of the world's rocks. Strata, they asserted could only be studied in situ, and fossils lost much of their scientific value if their location—both geologic and stratigraphic—was unknown. Geological research thus depended on field surveys and collecting trips. Because British geologists conducted their fieldwork on foot (often walking twenty miles in a day) and in exposed areas, the field season lasted from late spring until early fall. Curiously, the Geological Society modeled its summer recess on that of Parliament and the older learning societies and so did not suspend operations until several weeks after the field season began. Most leading geologists skipped the last few meetings of each year in order to make an early start on the fieldwork, which would provide the raw material for papers written during the upcoming year.[12]

Victorian geologists believed that fieldwork was a practical necessity but also saw deeper significance in it. The gentlemanly leaders of the geological community equated fieldwork with the "manly," chivalric qualities that they valued most: stamina, native intelligence, and high-spiritedness. Lyell, Murchison, and their colleagues styled themselves "brethren of the hammer, knights errant, a spiritual confraternity in search of a stratigraphic grail." For amateur enthusiasts in the community's rank-and-file, geological fieldwork was "rational recreation"—a chance to improve one's mind while enjoying an outing. Its appeal was similar to that of historical archaeology, and many local societies catered to students of both subjects. Finally, fieldwork's romantic aura appealed to geologists at all levels of the community. Doing fieldwork meant encountering nature at its most pristine and often at its most dramatic: in mountainous North Wales, along the desolate coast of Cornwall, and on the slopes of Sicily's Mount Etna. Such encounters gave geology a mystique that laboratory-bound sciences were hard pressed to match.[13]

12. M. J. S. Rudwick, *The Great Devonian Controversy* (1985), 37–41.
13. R. Porter, "Gentlemen and Geology," *Historical Journal* 21 (1978): 819–21.

Beyond its practical utility and its emotional appeal, field-work had an important philosophical role in British geology. Ge-ologists regarded detailed observations as the foundation of truth and so saw fieldwork as scientifically virtuous as well as spiritu-ally uplifting. The commitment to empiricism that they shared with other Victorian scientists led them to insist on strong links between fieldwork and theory. Appeals to "the rocks themselves" were the accepted method of settling disputes about theory or interpretation.[14] Data carried more weight if observed firsthand in the field than if gleaned from published papers, no matter how well documented those papers might be. An appeal to facts col-lected in the field was, in the context of theoretical disputes, an appeal to Nature itself.

The empirical methods that permeated British geology in the 1850s were part of a tradition stretching back to the beginning of the century. The early leaders of the Geological Society of Lon-don, founded in 1807, made such methods the basis of their re-search program. The society's early leaders appealed to strict em-piricism as a corrective to what they regarded as fruitless debate over abstract "theories of the Earth," insisting that geologists should describe Britain's strata and fossils in detail before propos-ing any interpretation of them. This attitude, though it softened somewhat by the end of the society's first decade,[15] set a pattern that lasted well into the 1860s. British geologists identified their science with stratigraphy and paleontology, which they defined primarily as taxonomic exercises. Defining the sequence in which strata or animals had appeared on Earth remained more im-portant to most Victorian geologists than formulating theories about the causes that brought them there.

British geologists agreed, as an article of faith, that the strata and fossils making up the Earth's crust told a clear, complete, unambiguous story. They believed that the geological chapters in the book of nature could be read by any observer willing to

14. M. J. S. Rudwick, "Hutton and Werner Compared," *BJHS* 1 (1963): 117–35. Secord, *Controversy;* Rudwick, *Great Devonian Controversy.*

15. R. Laudan, "Ideas and Organizations in British Geology," *Isis* 68 (1977): 527–38. M. J. S. Rudwick, "The Foundation of the Geological Society of London," *BJHS* 1 (1963): 325–55.

approach them with an unprejudiced mind. Theories were permissible, but only in their proper place—after extensive observation—and only if they took full account of the facts. This "self-conscious empiricism," to use James Secord's phrase, pervaded the articles and books written by the rank-and-file members of the geological community, and strongly influenced the theories proposed by its elite members.[16] Victorian geologists thus shared the historical archaeologists' belief that the past could be understood without resorting to theory. The two groups differed, however, in their degree of commitment to that belief.

Though almost as strongly committed to empiricism as their archaeological colleagues, British geologists espoused a less stringent form of the method. The historical archaeologists insisted that any attempt to systematize or generalize without first collecting enormous quantities of data was unscientific. They therefore rejected not only speculations on the origins of particular artifacts but also classificatory schemes such as the Three-Age System. Geologists took a similar attitude toward explanatory theories, but looked more favorably on classificatory ones. Though hesitant to speculate about the origins of a particular group of rocks, they felt free to hypothesize about how to classify and correlate the strata. Once proposed, these hypotheses were tested and refined through further fieldwork, debate at society meetings, and informal discussion in person or by post. This process of evaluating alternative stratigraphic systems was the basis of debates such as those chronicled by Martin Rudwick, James Secord, and David Oldroyd.[17]

Given the geological community's commitment to stratigraphy, its acceptance of classificatory theories was a virtual necessity.[18] Like William Buckland and his circle, whose primary concern was reconstructing ancient environments, British stratig-

16. M. J. S. Rudwick, "Cognitive Styles in Geology," in *Essays in the Sociology of Perception,* ed. M. Douglas (1982), 225–27.

17. Rudwick, *Great Devonian Controversy;* Secord, *Controversy;* D. R. Oldroyd, *The Highlands Controversy* (1990).

18. Rachel Laudan has suggested that the Geological Society's determination to be completely atheoretical in the years immediately after its founding significantly hindered the progress of British geology. See Laudan, "Ideas and Organizations."

raphers reveled in complexity. Buckland, however, hoped to use a complex collection of geological details to reconstruct an equally complex environment. The stratigraphers sought to reduce the bewildering variety of strata and fossils in the geologic record to a generalized—though still detailed—system. The stratigraphers believed that the geologic record, like the natural world as a whole, was intrinsically complex, but that its complexity masked a pattern. They saw it as their task to discover and amplify that system, much as Linnaeus had done for animals and plants.

To reduce the complexity of the geologic record to a universal system—even a complex one—the stratigraphers had to divide and subdivide the geologic past. With a hierarchical framework of eras, periods, and epochs in place and Europe's major strata fitted into it, geologists had a provisional system for determining the relative ages of newly described strata. By comparing fossils in the new strata to the groups of index fossils in the sections already defined, they could correlate the new stratum with strata in existing sequences. If the correlations were straightforward, then assigning the new stratum to an era, period, and epoch was simply a matter of sliding it into an existing temporal pigeonhole. If the new strata did not correlate clearly with existing sequences, however, the boundaries of epochs, periods, or even eras could be adjusted accordingly. Geologists regarded their willingness to make such adjustments—adjustments of theory to observation— as proof of their empiricist credentials. Provisional systems were essential to Victorian geologists' goal of reducing complexity to order. Without such systems to give it shape, stratigraphy would have remained a mass of local observations: interconnected, but lacking any element of universality.

The Geological Community

Victorian geology's empirical orientation and emphasis on fieldwork created a unique role for amateurs with local interests. Geology, like historical archaeology, appealed to amateur enthusiasts because it was a fundamentally democratic science. It required neither mathematical skill nor mastery of abstract principles, used no specialized equipment, and could be pursued anywhere. Equally important, it offered its participants clear and attainable goals. Amateur geologists, by carefully studying the rocks and fossils of their home county, could reconstruct the county's geological history much as historical archaeologists illuminated its written history. Local societies often catered to both geologists and archaeologists; indeed, at the local level, the two groups often overlapped.

Amateurs with local interests formed the bulk of both the geological and archaeological communities, but they played very different roles within those communities. Leading historical archaeologists, because of their concern with the particular, saw research on local history as an end in itself. Leading geologists, because of their interest in international and intercontinental correlations, saw research on local geology as part of a larger program. Amateur geologists thus led a dual intellectual life. Their work satisfied their own curiosity about local rocks but also contributed to the larger research program carried on by the geological elite.

Stratigraphic generalizations, as well as explanatory theories, rested on detailed local knowledge that could only be gathered in the field. Fieldwork was time-consuming, however, and the field season comparatively short. Leading geologists such as Lyell, Murchison, and Sedgwick therefore relied on the work of hundreds of amateur geologists scattered throughout the British Isles. The best local amateurs became acknowledged experts on the rocks of their home counties, to which they had often devoted years of study. They shared their expertise with the London-based elite in published papers, in scores of letters, and sometimes in person. Geologists doing fieldwork in unfamiliar areas often sought guidance from local amateurs on the fine points of local stratigraphy and locations of the best outcrops.

J. W. Salter, chief paleontologist to the Geological Survey, stressed the importance of amateur geologists in an 1858 letter to the amateur-oriented magazine *The Geologist*. The key to the progress of geology, Salter wrote, "is *abundance of good facts*, and these can only be collected by the industry of local observers who will communicate these results in a tangible form, available for the workers on particular subjects."[19] Salter referred to strati-

19. J. W. Salter, letter, *Geologist* 1 (1858): 301.

graphic subjects, but amateur contributions of facts were equally important in resolving major theoretical issues. Debates about changing land and sea levels, for example, depended on amateurs' observations of raised beaches and sunken forests near their homes.[20] Amateurs played a similar role in debates about the extent of the glacial period by reporting the presence of ice-scratched rocks, large boulders, and other glacial phenomena in their localities.

Because the issue of professionalization looms so large in the historiography of nineteenth-century science, it is important to define the term *amateur* with some precision here. A professional can be defined as someone who possesses an agreed-upon set of qualifications and is paid for his or her work. Applied to the British geological community in the mid-nineteenth century, however, this definition obscures more than it clarifies. In the 1850s it was possible, though not easy, to make a living as a paid geologist in Britain. Sedgwick, for example, held the professorship of geology at Cambridge; Andrew Ramsay, J. B. Jukes, and many less talented men worked for the Geological Survey; and Henry Keeping worked as a paid fossil collector. Though all these men were paid for their geological work, they had little else in common, sharing neither a common career path nor a common intellectual outlook. To call them professionals is to impart to them a social and intellectual cohesiveness that they did not possess. Sedgwick, for example, had far more in common with his financially independent coworker Roderick Murchison than he did with a working-class collector like Keeping or a junior member of the survey.

Roy Porter's term *career geologist* is significantly more useful than *professional geologist* when discussing the mid-Victorian geological community. Though the differences between paid and unpaid geologists were often slight, a vast gulf separated those who worked at geology full time and those who pursued it only as a recreation.[21] If they were financially independent, like Murchison or Charles Darwin, career geologists were free to devote the ma-

20. Based on a survey of papers published in the *Quarterly Journal* of the Geological Society and the *Report* of the British Association during the 1850s.
21. Porter, "Gentlemen and Geology," 811–16.

jority of their time to geology—reading, correspondence, meetings, and fieldwork in the summer. If not, they sought jobs—like those on the Geological Survey—that would aid rather than hinder their research. Lyell, for example, first abandoned his legal career and then resigned a professorship of geology at King's College, London, finally supporting himself by writing, and trading financial security for more time to pursue his research.[22] Henry De la Beche, Ramsay, Jukes, and other members of the survey put up with low pay and an austere, bachelor existence for years rather than resign their posts and have their research opportunities curtailed. These "career geologists" became the elite of the British geological community because they alone had the time to master the masses of details that underlay high-level theory.

The crucial division in the British geological community, then, was not between amateurs and professionals but between amateurs and career men. In the remainder of my discussions, I will use *amateur geologist* to mean one who—by choice or because of financial constraints—pursued geology only part time, and whose work was limited in scope and impact as a result. There are, inevitably, exceptions to this rule who defy any attempt to define them into conformity with it. The most glaring exception is Joseph Prestwich, who worked full-time as a London wine merchant but was, nonetheless, a leading stratigrapher and a key figure in the investigation of human antiquity.

Career geologists saw their amateur colleagues as competent scientists and full-fledged members of the geological community. Amateurs attended meetings of the Geological Society, presented papers, and participated in debate. Tacit assumptions about the extent of their competence limited the amateurs' participation, however. Career geologists routinely accepted their amateur colleagues' factual statements about the geology of their local areas but dismissed the amateurs' opinions on theoretical questions. Questions of method, theory, or interpretation were the province

22. For Lyell's career, see R. Porter, "Charles Lyell," *Janus* 69 (1982): 29–50; for Murchison's, see J. A. Secord, "King of Siluria," *Victorian Studies* 25 (1982): 413–42. For an interesting account of a geological career that ultimately failed, see J. A. Secord, "J. W. Salter," in *From Linnaeus to Darwin*, ed. A. Wheeler and J. Price (1985).

of career geologists. Members of the omnicompetent geological elite such as Lyell and Murchison, could pronounce on any question and be listened to with respect. Career geologists like Prestwich, whose expertise was more specialized, played a somewhat more circumscribed role. Their opinions on questions regarding their specialty carried great weight; even members of the elite deferred to them on technical issues. Their opinions on other questions carried little weight unless corroborated by specialists in the relevant field.[23]

Smaller groups, which Rudwick calls "core sets," formed and reformed within the ranks of the career geologists. Core sets formed in response to the emergence of "focal problems"—pressing theoretical problems that stood out against the background work of systematizing the world's strata. The problems themselves were sometimes stratigraphic, such as the "Great Devonian Controversy" of the 1830s. More frequently, the focal problems involved questions about the causes, rate, and extent of geologic change—questions that lay outside everyday geological practice. Core sets consisted of geologists whose specialties directly related to a given problem. Their members investigated the problem and debated possible solutions, consulting with amateurs and outside specialists as the need arose.

Core sets were transient, but extremely influential, bodies. Because a core set consisted, by definition, of those most qualified to solve a given problem, solutions formulated by core sets were routinely accepted by the rest of the community.[24] Once it had found a solution to its problem, a core set disbanded and its members turned to everyday stratigraphic work or to other focal problems related to their specialties. Most core sets left no formal published record of their work; their members can only be identified through their individual publications and the content of their correspondence. It is vital, however, to identify such core sets and to evaluate their contributions. Victorian geology progressed less

23. Rudwick, *Great Devonian Controversy*, 418–25, and "Charles Darwin in London," *Isis* 73 (1982): 186–206.

24. Rudwick, *Great Devonian Controversy*, 426–28; for a general discussion of "core sets," see H. M. Collins, "Place of the Core Set in Modern Science," *History of Science* 19 (1981): 6–19.

through the work of individuals than through the efforts of individuals in concert.

Geologists and Human Antiquity

Geologists and archaeologists agreed, in the years before 1858, that the first appearance of humans was a major event in the history of the Earth. It divided the modern world in which humans lived from a series of former worlds populated by now-extinct animals. The advent of humankind also marked the intellectual frontier between geology and archaeology. It signaled the end of the Post-Pliocene epoch and thus of geologic time. Geologists might study ongoing geological processes or events such as volcanic eruptions that took place after the appearance of humans, but they left humankind to the archaeologists. In principle, geologists and archaeologists should have been equally interested in the age of the human race, but in practice they were not. The British archaeological community took virtually no notice of the human antiquity question between 1800 and 1858.[25] Its limits were defined and evidence bearing on it was evaluated solely by members of the geological community.

The historical archaeologists' ideas about the pre-Roman history of Britain account for much of this striking dichotomy. The leaders of the archaeological community regarded the pre-Roman, or Celtic, period as all but opaque to scientific methods because of its lack of written records and paucity of artifacts. They held out little hope that life in Celtic Britain might someday be reconstructed as life in Roman or medieval Britain had been. The historical archaeologists' close ties to history reinforced their lack of interest in the pre-Roman period. It was too far from and too unlike the present to shed much light on the development of the British people and nation. The first stages of the modern era, when humans were new to the Earth, were times that historical archaeologists were neither equipped to investigate nor interested in studying.

Geologists, on the other hand, had both strong motives and

25. The few interpretive discussions that did appear in national archaeology journals explicitly followed the geologists' arguments. See, e.g., T. W. Smart, "Account of Some Ancient British Antiquities," *JBAA* 2 (1845): 171–74.

well-tested methods for resolving the question of human antiquity. The overarching goal of early Victorian geology was to systematize the world's rocks by imposing a chronological classification on them. The appearance of humankind marked the upper boundary of the last geologic period and so of the entire classification system. Fixing the moment at which humans first appeared was, therefore, a problem of considerable importance. It was also a comparatively straightforward problem. Establishing the age of human remains—bones and artifacts—was a matter of determining their stratigraphic relationship to nearby animal bones. British geologists, who saw stratigraphy and paleontology as the heart of their science, assessed the relative positions of fossil animal bones on an almost daily basis.

It was geologists, therefore, who passed judgement on the many claims made between 1820 and 1850 for the antiquity of humankind. Donald Grayson and other historians have analyzed these claims and the excavations that provided evidence for them in detail. It will suffice here to view those that attracted the most attention from contemporary observers and to highlight the British geological community's reaction to them.

William Buckland believed that caves containing rich deposits of bones provided important information about extinct faunas. His 1820 excavation of one such cave—Goat Hole, near Paviland on the Welsh coast—uncovered not only animal bones but also a human skeleton. The skeleton, dubbed the "Red Lady of Paviland" because of red ochre stains on the bones, lay alongside the bones of mammoths and other extinct animals. Buckland argued that this association of human and extinct animal remains was accidental: during the Roman occupation of Britain, the Red Lady had been buried in a grave that happened to reach down into the stratum containing mammoth bones. The complexity of cave strata, he cautioned, meant that such stratigraphic associations had to be examined with great care.[26]

Buckland used similar arguments to explain the discovery of stone tools among extinct animal bones in Kent's Cavern, near

26. Rupke, *Great Chain of History*, 89–95; D. K. Grayson, *The Establishment of Human Antiquity* (1983), 65–69; F. J. North, "Paviland Cave," *Annals of Science* 5 (1942): 91–128.

Torquay, in Devonshire. Kent's Cavern was an extensive cave whose limestone floor concealed a layer of earth studded with bones of extinct animals. Father John MacEnery, a local naturalist, dug through the floor and into the earth layer below in a series of excavations in 1825, 1826, and 1829. Above the bones, he discovered artifacts from the Roman era; among the bones, he discovered primitive stone tools. No expert in cave paleontology, MacEnery asked Buckland for his opinion of the discoveries and was told that, while old, the stone tools were younger than the bones around them. Buckland suggested that the tools had been dropped into oven pits dug by the cave's Celtic inhabitants and subsequently sealed up by accumulating limestone.[27]

MacEnery split the difference between his original belief in the tools' great antiquity and Buckland's belief in their recency, concluding that the extinct animals in Kent's Cavern predated the biblical deluge, while the stone tools postdated it slightly. This, he pointed out, would make the first Britons older than Buckland and other leading scientists had estimated.[28] MacEnery left Torquay and his cave research in 1830. The manuscript of MacEnery's book—still unfinished at his death in 1841 and sold as waste paper in 1842—was not published until 1859, when it was obsolete.

The most detailed early studies of the human antiquity problem were done on the Continent, but British geologists paid close attention to them. Jules de Christol, exploring caves near Montpellier, France, in the 1820s, discovered pottery and human bones in strata that also contained the remains of extinct hyenas, rhinoceroses, and bears. On the basis of Christol's finds and his own 1828 discoveries in caves near Bize, Paul Tournal argued that humans and extinct animals had once coexisted in France. Phillipe-Charles Schmerling's discoveries of two human skulls, an assortment of other human bones, and several flint tools in a pair of caves near Liège, Belgium, corroborated Tournal's theory.

27. Grayson, *Human Antiquity*, 72–76; A. S. Kennard, "The Early Digs in Kent's Hole," *Proceedings of the Geologists' Association* 56 (1945): 156–213. Kennard's paper includes an invaluable guide to the surviving portions of MacEnery's "lost" manuscript.

28. J. MacEnery, *Cavern Researches*, ed. Edward Vivian (1859), 65.

The human remains, mingled with rhinoceros and hyena bones, convinced Schmerling that humankind had coexisted with now-extinct animals long before the Earth had assumed its "modern" form.[29]

Lyell, Buckland, Gideon Mantell, and other leading British geologists read the reports of these developments but were not convinced. Buckland visited the caves of southern France as well as those in Belgium but dissented sharply from Tournal's and Schmerling's conclusions in his *Geology and Mineralogy* (1836). He maintained that associations between human remains and extinct animals on the Continent were, like those in Kent's Cavern, accidental. Lyell, though impressed by a visit to Liège in 1833, argued in *Principles of Geology* that Schmerling's finds were insufficient to prove the great antiquity of the human race. Like Buckland and French geologist Jules Desnoyers, Lyell believed that cave stratigraphy was too complex to be taken at face value.

The human antiquity question attracted little attention from the mid-1830s to the mid-1840s, but came to the fore again in 1846–47 when amateur geologists Edward Vivian and William Pengelly carried out a new excavation of Kent's Cavern. The work convinced both Vivian and Pengelly that MacEnery's claims about human antiquity were correct but failed to gain them a hearing before the Geological Society of London. The society, unsympathetic to Vivian's mingling of claims about human antiquity with claims about the geological significance of Noah's flood, rejected his paper on the Kent's Cavern excavation. Two years later, in 1849, the first volume of Jacques Boucher de Perthes's *Celtic and Antediluvian Antiquities* achieved wide circulation. De Perthes reported that he had dug for ten years in gravel beds along the Somme River near Abbeville, France, uncovering dozens of stone artifacts mingled with remains of extinct animals in strata several meters below the surface.[30] The Somme gravels were stratigraphically simple, and the artifact-bearing stratum was too deep to have been disturbed by graves or oven pits, but British geologists dismissed De Perthes's work as "rubbish."[31]

29. Grayson, *Human Antiquity*, 99–112.
30. No biography of De Perthes exists in English, but see C. Cohen and J.-J. Hublin, *Boucher de Perthes* (1989).
31. Grayson, *Human Antiquity*, 117–26.

In this litany of discovery and rejection, three trends deserve emphasis. First, as early as the 1820s, leading British geologists recognized the age of the human race as a geological question. Rather than consult leading archaeologists, they took it upon themselves to investigate and pass judgement on each new set of data that was presented as evidence for human antiquity. Second, though they clearly recognized the religious and philosophical implications of the human antiquity problem, geologists couched their discussions in stratigraphic and paleontological terms. Their rejection of many arguments for human antiquity stemmed, as Grayson has shown, from the ambiguity of the stratigraphic and paleontological data involved.

Finally, British geologists' responses to the evidence for human antiquity reflected both the structure and the methodological commitments of the geological community. The men who argued for human antiquity between 1820 and 1850 were, at best, amateur geologists. Schmerling paid little or no attention to complex stratigraphy of the caves he excavated.[32] Many others— MacEnery, Tournal, and de Perthes, for example—had done no geological fieldwork at all before they began the excavations that yielded human remains. The amateurs' narrowly circumscribed experience limited their credibility in the eyes of career geologists. Lyell and Buckland never questioned the amateurs' facts but had serious doubts about their conclusions. Proposed by a similarly inexperienced group of amateur geologists, any such radical revision of an established theory would have met a similar reception.

The unorthodox quality of many of the amateurs' theories strengthened the career geologists' suspicion of the theories. Boucher de Perthes argued in the first volume of *Celtic and Antediluvian Antiquities* (1849) that a series of worldwide catastrophes shaped the Earth, each destroying a complete fauna that included humans. De Perthes's identification of the last catastrophe with Noah's flood was, in the eyes of British geologists, only one of his sins.[33] Yoked to such an outdated picture of Earth history, his claims about human antiquity were literally meaningless.

32. S. J. De Laet, "Philippe-Charles Schmerling," in *Towards a History of Archaeology*, ed. G. Daniel (1981), 115–16.
33. Grayson, *Human Antiquity*, 128–30.

Vivian's similar reliance on outdated theories provides the most likely explanation of the Geological Society's refusal to publish his 1847 Kent's Cavern paper. He was committed to reconciling scripture with the facts of geology—a goal shared by many amateur geologists but, by the late 1840s, virtually no career geologists. Vivian's later work on human antiquity demonstrated his firm belief that the humans and extinct animals found together in cave deposits were casualties of Noah's flood. By the late 1840s, however, the leaders of the geological community had long since ceased to regard Noah's flood as a geological agent.[34] The 1847 paper on Kent's Cavern was Vivian's work, and if it was filled with what Pengelly later called "geo-theological speculations," the Geological Society would certainly have dismissed it as unsound.

Methodological constraints further restricted the human antiquity advocates' impact on established geological ideas. Victorian geologists were empiricists, but to varying degrees. United by their belief that any theory should rest on a foundation of carefully observed facts, they differed on the quality and quantity of facts required for such a foundation. Discussions of the human antiquity problem between 1820 and 1850 reflected these differences. Some geologists, like Vivian and Pengelly, believed that the evidence from Torquay, Liège, and Abbeville was sufficient to establish the coexistence of humans and extinct animals. Others, who cited the cautionary statements of Buckland and Desnoyers, did not believe the evidence to be sufficient.

The prevalence of skepticism on the human antiquity question reflected many things: the minimal reputations of the most outspoken human antiquity advocates, the complexity of cave stratigraphy, and the disturbing philosophical implications of men among the mammoths. It also reflected the fact that revising estimates of the human race's age meant revising the boundaries of the Quaternary period and, ultimately, of geology itself. Revisions of that magnitude demanded a broad foundation of indisputable evidence. Many amateurs and nearly all career geologists believed, in 1850, that such a foundation did not exist.

34. Vivian, "Human Remains in Kent's Cavern," *BAAS Report* 26 (1856): 119–20; J. R. Moore, "Geologists and Interpreters of Genesis," in *God and Nature*, ed. D. Lindberg and R. Numbers (1986).

During the 1850s, however, changes in geological theory set the stage for a new consideration of human antiquity. They did not, in themselves, change anyone's mind about the age of the human race, but they made human antiquity more plausible. They ensured that the geological community would be acutely interested in the new evidence for human antiquity discovered late in the decade. The changes emerged from studies of three distinct problems: progression in the fossil record, Quaternary stratigraphy, and the glacial theory. Many geologists who did important work on these three problems became leading figures in the renewed investigations of human antiquity that began in 1858.

The idea that life on Earth had grown steadily more complex over time had become a firmly established part of paleontology during the early eighteenth century. The extensive paleontological work of the early nineteenth century reinforced the idea of progression in the history of life by providing empirical support from the fossil record. In Britain, progressionism became tightly intertwined with the directionalist view of Earth history advocated by most leading geologists. The idea of progression in the fossil record also supported the tradition of natural theology popularized by William Paley at the beginning of the nineteenth century and reinforced in the Bridgewater Treatises of the 1830s. Progressionism became an article of faith among natural theologians that both the history of the Earth and the history of life represented the unfolding of a divine plan designed to produce an Earth perfectly suited for human habitation. Progressionists argued that the fossil record showed the advent of increasingly complex classes of animals, and of increasingly complex animals within each class. Thus, the fossil record reflected not only progress toward humanity but also the providential design that suited new species to an increasingly complex environment.[35]

Adam Sedgwick alluded to progressionism in his 1831 presidential address to the Geological Society, and William Buckland explicitly promoted it in his 1836 book *Geology and Mineralogy.*[36] Charles Lyell, committed to a steady-state view of both the his-

---

35. P. J. Bowler, *Fossils and Progress* (1975), 1–15.
36. A. Sedgwick, "Address of the President," *Proceedings of the Geological Society of London* 1 (1820): 281–316; W. Buckland, *Geology and Mineralogy* (1836). For

tory of the Earth and the history of life, rejected progressionism in the first (1830–33) edition of his *Principles of Geology* and defended that position for nearly three decades. Lyell argued that the progression visible in the fossil record was an illusion—a result of the record's imperfection. He supported this position with a variety of paleontological observations. The earliest fossil plants, Lyell claimed, included some with high levels of organization, as did the earliest invertebrates. Reptilian footprints had been discovered in Carboniferous rocks, supposed avian footprints in Triassic rocks, and mammalian bones in Jurassic rocks, all suggesting that these classes of vertebrates had appeared much earlier than most paleontologists supposed.

Lyell intended his litany of paleontological anomalies to show that all the major classes of animals had existed side by side for most of Earth's history. Individual species had been been created and become extinct, but the overall complexity of life on Earth had changed very little over many millions of years. He agreed that humans had arrived on Earth only recently but denied that they represented the endpoint of any geological or biological progression. Lyell's nonprogressionism remained a minority position among British geologists, though it won over such noted paleontologists as Gideon Mantell, Edward Forbes, and eventually T. H. Huxley.[37]

The debate over progression was resolved, during the 1850s by the emergence of a new interpretation of the fossil record. Drawing on a series of recent paleontological studies, Richard Owen argued that neither Lyell's nonprogressionism nor the old, unilinear progressionism was a tenable theory. As well as showing that many of Lyell's paleontological anomalies had been misidentified or stratigraphically misplaced, Owen pointed out that the increasingly complete fossil record offered increasingly compelling evidence of progression. Owen also demonstrated, however, that progress within any given class involved many concurrent lines of development. In doing so, he equated progression

---

more on the geological ideas of Sedgwick, Buckland, and their intellectual circle, see Rupke, *Great Chain of History*, 109–200.

37. Bowler, *Fossils and Progress*, 69–79.

with divergence. The reptiles, for example, had begun as a few simple, generalized species and progressed to become hundreds of more complex, more specialized species.[38]

By undercutting the idea that the fossil record showed an essentially unlinear progression, Owen's work subtly changed the face of natural theology. Earlier natural theologians had seen the fossil record as evidence of a single series of divinely planned geological and biological changes that had prepared the Earth for the arrival of humans. The natural theologians of the 1850s, on the other hand, saw the fossil record primarily as evidence of how each class had grown gradually more diverse over time, and fulfilled the adaptive potential that God had given it.[39] The shift from linear progression to progression with divergence significantly altered humankind's position in the progressionist view of Earth history. Humans had been seen as the culmination and raison d'être of the entire history of life. They became, for those who accepted the new model, the culmination of one among several parallel lines of progression. The beginning of the modern world thus became more difficult to define. It was still possible to equate that beginning with the advent of humankind, but such an equation could no longer be regarded as self-evident.

The redefinition of progressionism did not, in itself, render the antiquity of the human race any more plausible. Richard Owen, the architect of the new progressionism, was highly skeptical of the evidence for human antiquity. Progression with divergence did, however, make the appearance of humans on an unfinished Earth more plausible by making the two issues independent. It thus made human antiquity significantly more compatible with a progressionist view of the fossil record. The redefinition of progressionism lowered what had been a long-standing barrier to belief in human antiquity, facilitating scientists' acceptance of human antiquity on the basis of new evidence.

British geologists' investigation of comparatively recent rocks during the 1850s provided a context for interpreting the new evidence. The British geological community's attempt to systematize

38. Ibid., 99–107.
39. Ibid., 104–106.

the world's rocks was nearly complete by the 1850s. The broad outlines of the system were almost finished, and only the very oldest and very youngest fossil-bearing rocks remained to be put into their places. The oldest rocks, which Adam Sedgwick designated the Cambrian system and Roderick Murchison subsumed under his Silurian system, became the center of a bitter, thirty-year controversy.[40] The strata of the geologically recent Quaternary system, though not as controversial, proved just as difficult to reduce to a workable system.

The Quaternary deposits posed several unique stratigraphic problems. First, the Quaternary strata had been deposited over a relatively short time and under rapidly changing conditions. As a result, they were much thinner and more difficult to correlate than the strata of older periods. Second, they commonly existed only in small, widely separated patches. The strata from earlier periods could often be traced for miles in seaside cliffs and cuttings for new railway lines, but Quaternary strata generally had to be sought out in dozens of isolated basins, caves, and river valleys. Finally, many of the animals whose fossils were found in the Quaternary strata of Europe belonged to extant European species or their close relatives. This reduced the number of reliable index fossils—which, by definition, had to belong to extinct species—and put a premium on often subtle distinctions between the remains of extant and extinct species.

The Quaternary strata that most concerned British geologists were those of southern England and western France—a series of thin, sometimes interleaved beds of clay, gravel, and loam. In southwest England, their stratigraphy was worked out by R. A. C. Godwin-Austen—an accomplished geologist who had also carried out excavations in Kent's Cavern during the early 1840s.[41] Godwin-Austen's colleague, Joseph Prestwich, made a specialty of the strata of southeastern England, Belgium, and western France. Prestwich devoted thirteen years and nearly thirty papers

40. The controversy came to a head in 1852, when the Geological Survey colored its maps of Wales according to Murchison's system. The geological community reached a compromise solution in 1879, though Murchison's protegé, Archibald Geikie, was still grumbling about the compromise as late as 1905. Secord provides a detailed narrative and analysis in *Controversy in Victorian Geology.*

41. H. B. Woodward, "Robert Alfred Cloyne Godwin-Austen," *Geological Magazine*, n.s., 2 (1885): 1–10.

to the Quaternary and somewhat older Tertiary strata on both sides of the English Channel. His papers, published in the Geological Society's *Quarterly Journal*, were models of well-documented caution. Examining one small group of strata at a time, they outlined stratigraphic relationships, fossil contents, and, finally, correlations.[42] By 1858, Godwin-Austen and Prestwich had reduced the complex, recent strata around the English Channel to an intricate but comprehensible system. In doing so, they defined the stratigraphic stage on which the investigations of human antiquity would be played out.

One of the most intriguing focal problems of the 1850s turned on a group of unconsolidated deposits, collectively called the Drift, that lay above the rocks of the Pliocene epoch and below the Alluvium, which had been deposited by running water in the very recent past. In his 1852 presidential address to the Geological Society, William Hopkins took the Drift as his primary subject, stating that it was laid down in "a period of peculiar conditions, and of phaenomena referable to peculiar causes, the study of which has opened to us entirely new views respecting the agencies which have so marvelously modified the face of our planet, by the continual transfer of matter from one part of its surface to another." The agencies to which Hopkins referred were valley glaciers, drifting icebergs, and marine inundations. All three, he concluded, had played roles in spreading the Drift deposits—including the enormous boulders known as "erratics"—over northern Europe.[43] Like virtually all British geologists of the 1850s, Hopkins dismissed Louis Agassiz's theory of an ice age during which continentwide glaciers spread over Europe and left the Drift in their wake.

During the 1850s, most British geologists followed Hopkins in attributing the Drift to a mixture of causes, though they disagreed about the relative importance of each.[44] The evidence that would resolve the problem lay in the Drift itself. Geologists frequently cited the presence of marine shells in a particular locality as proof of a recent inundation. Fossils from species now confined

---

42. G. Prestwich, *Life and Letters of Sir Joseph Prestwich* (1899), 63–110; for a list of the papers, see 423–25.

43. W. Hopkins, "Address of the President," *QJGSL* 8 (1852): xxv–xxviii.

44. G. L. Davies, *The Earth in Decay* (1969), 275–98.

to cold climates, on the other hand, were cited as evidence of formerly colder temperatures that could be tied to advancing valley glaciers. Ancient beaches, far above modern sea level, offered strong evidence that a marine inundation might once have taken place. Drowned forests suggested that glaciers had once held enough water to significantly lower the level of the sea. The efforts of the leading members of the geological community—notably Lyell, Ramsay, Prestwich, and Hopkins—to establish the origins of the Drift gave the community an intimate acquaintance with the specific deposits in which the evidence for human antiquity lay.

Despite their growing familiarity with Quaternary deposits, British geologists published virtually nothing on human antiquity between 1850 and 1858. The comments that did see print were, with the exception of a single article by Gideon Mantell, confined to brief passages in textbooks and popular works on geology. Geologists offered a spectrum of opinions on the age of the human race, but both their endorsements and their dismissals of the evidence for human antiquity were cautious and restrained. The geological community seemed content to wait for more and better evidence before passing final judgment on the idea of human antiquity. If their caution was simply a rhetorical disguise for dogmatism, then it was an uncommonly effective disguise.

In the fifth (1855) edition of his *Manual of Elementary Geology*, Charles Lyell grouped the Earth's most recent strata under the term *Post-Pliocene*. Following his principle, already well established for the Tertiary Period, of defining epochs according to the ratio of extant to extinct mollusc species present, Lyell defined the Post-Pliocene epoch as that in which all the fossil shells belonged to still-living species. He assigned to it both the alluvial deposits that "can be shown to have originated since the Earth was inhabited by man," and the Drift, "in which no sign of man or his works can be detected."[45] While Lyell apparently rejected the validity of the idea that humans had existed in the Drift period, he carefully made his rejection conditional. He did not claim

45. C. Lyell, *A Manual of Elementary Geology*, 5th ed. (1855), 117–18. For other leading geologists' statements that the age of the human race was an open question, see L. Horner, "Address of the President," *QJGSL* 3 (1847): xxxvi; and H. T. De la Beche, *A Geological Manual* (1832), 173–74.

categorically that humans had not lived during the Drift's formation, but only that no traces of them had yet been found in it. A
master rhetorician like Lyell would have been fully conscious of
the semantic distance separating the two statements.

In his own *Manual of Geology*, published in the same year as
Lyell's fifth edition, John Phillips took a similarly cautious position on human antiquity. Phillips briefly surveyed the well-
established arguments for and against the coexistence of humans
and extinct mammals. He cited the work of Tournal and Christol
as strong evidence for human antiquity but pointed out the problems posed by secondary burials. He noted that almost no human
bones had been found in the European parts of the Drift—a point
against human antiquity—but suggested that more, and earlier,
remains might be found in central Asia, where the human race
arose. Phillips characterized the available evidence concerning
the coexistence of humans with mammoths and extinct carni-
vores as "very imperfect" but concluded that "it is not, perhaps,
an unreasonable expectation that, eventually, this question will
be decided in the affirmative."[46]

Henry De la Beche, then director of the Geological Survey,
also considered the evidence for human antiquity in print. He
discussed the complexity of cave stratigraphy in his *Geological Observer*, pointing out that the remains of extant species were frequently found mingled with those of extinct ones. Caution, De la
Beche reminded his readers, was essential. Scientists should neglect neither the presence of human remains among the bones of
extinct animals nor "the accidents which may have brought such
apparently contemporaneous mixtures together." De la Beche
cited Buckland's work on the "Red Lady of Paviland" as proof
that caves required careful investigation. He noted, skeptically,
Schmerling's descriptions of human fossils found in caverns near
Liège. The subject of human antiquity, he concluded, "is one of
no slight interest, and requires at least very careful investigation,
without prejudgement of any kind."[47]

Gideon Mantell, a noted paleontologist, discussed human antiquity at length in an 1850 address to the Archaeological Insti-

46. Phillips, *Manual of Geology*, 435–38.
47. H. T. De la Beche, *The Geological Observer*, 2d ed. (1853), 301–4; quota-
tions on 301, 304.

tute on the "Connexion between Archaeology and Geology." His speech, reprinted in the *Edinburgh New Philosophical Journal*, was the only extended treatment of the human antiquity problem published by a British geologist between 1850 and late 1858. Mantell dismissed most of the evidence recently offered as support for human antiquity. He rejected the evidence from the Somme Valley because Boucher de Perthes had clearly mistaken naturally chipped flints for stone tools, and that from Kent's Cavern and similar sites because disturbances of the cave deposits after the deposition of the tools made accurate stratigraphy virtually impossible. Despite this, Mantell was convinced of the coexistence of humans and extinct Post-Pliocene mammals. He cited a case in which a human skeleton, wrapped in a deerskin, had been found in an Irish peat bog alongside the skeleton of an Irish Elk, one rib of which had been pierced by a sharp object when the animal was still alive.[48]

This evidence, Mantell argued, conclusively proved that humans existed "at that remote period when the Irish Elk, and other extinct species and genera of terrestrial mammalia, whose remains occur in the superficial alluvial deposits, inhabited the countries of Europe." More important, because the Irish elk was known to have been contemporary with the mammoth, hyena, and cave bears, it was "not improbable that sooner or later human remains may be discovered coeval with the bones of those animals."[49] Thus, Mantell was willing to accept the existence of humans during the Drift period, even in the absence of any direct physical evidence for it. Logic dictated their presence, and the chances of finding confirmatory evidence seemed good. The 1854 edition of *Medals of Creation*, Mantell's popular manual of paleontology, suggested that such evidence had indeed been found—though he treated it with characteristic caution. Discussing fossilized human skulls found alongside elephant bones in the Alps, Mantell concluded that these facts, "if correctly reported, naturally lead to the conclusion that human beings were contemporaneous with the extinct elephants . . . in the region referred to."[50]

48. G. Mantell, "Illustrations of the Connexion between Archaeology and Geology," *Edinburgh New Philosophical Journal* 50 (1851): 235–54.
49. Mantell, "Connexion," 252.
50. Mantell, *Medals of Creation*, 2d ed. (1854), 2:815.

Mantell's acceptance of human antiquity, Lyell's conditional denial, De la Beche's skeptical interest, and Phillips's hesitant optimism were all the products of a shared approach to geology. All four writers saw serious flaws in the bulk of the evidence advanced to support human antiquity. Even Mantell rejected the now-famous evidence from Abbeville, Kent's Cavern, and other such sites in favor of a relatively obscure find in Ireland. None of them, however, was willing to dismiss on general principles the possibility that humans had coexisted with extinct animals. Phillips and Mantell fully expected that further research would produce conclusive evidence of coexistence. De la Beche insisted on detailed research and high standards for what counted as evidence, but emphasized the need for open-minded investigation. Though Lyell regarded the discovery of men among the mammoths as unlikely, he went out of his way not to rule out the possibility.

By the mid-1850s, the decline of Mosaic geology and the modification of natural theology had lowered the principal philosophical barriers to a belief in human antiquity. At the same time, British geologists working on focal problems such as Quaternary stratigraphy and the origins of the Drift had become intimately familiar with the geological context of what were alleged to be the earliest human remains. This new familiarity reinforced the skepticism that most geologists had felt in the 1830s and 1840s but also bred the cautious optimism evident in Mantell's and Phillips's work. British geologists were more aware than ever of the complexity of sites such as Kent's Cavern but also more confident than ever that the human antiquity problem could be solved.[51]

The British geological community reserved judgment on the human antiquity problem as the 1850s drew to a close. Following the lead of Phillips, Mantell, and perhaps even Lyell, it waited for new and unambiguous evidence. When such evidence was discovered in the spring of 1858, British geologists were well prepared to appreciate and take advantage of it.

51. Cf. Grayson, *Human Antiquity*, 83–86; G. Daniel, *150 Years of Archaeology* (1975), 40–56; and J. W. Gruber, "Brixham Cave and the Antiquity of Man," in *Context and Meaning in Cultural Anthropology*, ed. M. Spiro (1965).

FOUR

Solving a Geological Problem

Discussions of the human race's age and humankind's place in the fossil record were common but sporadic during the first half of the nineteenth century. They engaged, at one time or another, the attention of most of Britain's leading geologists. These discussions did not, however, add up to a sustained investigation of the human antiquity problem; the geological community considered new claims for human antiquity on a case-by-case basis, without taking a formal position on the age of the human race. Until 1858, British geologists' informal consensus on human antiquity was that it was possible but as yet unproven.

The geological community's first concerted investigation of the human antiquity problem began in the summer of 1858, during the excavation of a newly discovered cave in southwest England. By the following summer, the investigation encompassed data from sites throughout Western Europe. The principal investigators presented this data—and drew literally epoch-making conclusions from it—in a series of papers read before Britain's leading scientific societies in the spring and summer of 1859. The papers discussed different sites but drew a single conclusion: humans had lived among now-extinct mammals in a Europe that was topographically and climatically different from that of the present day. Formulated by a small group of expert geologists, this new consensus on human antiquity was quickly adopted by the rest of the geological community. Charles Lyell spoke for virtually all British geologists when he declared his support for the new consensus in his 1859 address as president of the geological section of the British Association for the Advancement of Science. The investigations that transformed once-tenuous speculations into well-supported theories took less than eighteen months; they caused, a contemporary observer remarked, "a great and sudden revolution" in ideas about the past.[1]

The 1859 papers that announced the new consensus provided more than just descriptions of sites and artifacts. They presented a brief, carefully constructed history of how the investigators—that is, the papers' authors—had gathered their data and drawn their startling conclusions. According to this history, the new consensus had sprung, fully formed, directly from the data. Each author described how he had gone to one of the principal sites, confronted the facts, and accepted the presence of humans among the extinct animals of the Post-Pliocene epoch. The investigators' stories of conversion through fieldwork placed the investigators firmly in the tradition of Baconian empiricism, and so affirmed their methodological credentials.

These brief, sanitized tales of discovery form the core of nearly all recent histories of the establishment of human antiquity.[2] Historians have, therefore, overemphasized both the magnitude and the suddenness of the intellectual changes wrought in 1858 and 1859. The unpublished correspondence and papers of the principal investigators confirm that geologists' work on the human antiquity problem emerged from, and blended seamlessly with, their ongoing program of stratigraphic and paleontological research. The same archival material reveals that the new consensus emerged not from the data but from an intensive process of fieldwork and debate among a committee of expert geologists. The geologists who presented their conclusions as a fait accompli in the spring and summer of 1859 had disagreed sharply over them for much of the preceding year. That they did reach a consensus early in spring 1859 is due more to the intensity of their investigation than to the compelling nature of the data. The British geological community's new consensus on human antiquity

1. C. Murchison, "Editor's Note," in *Paleontological Memoirs and Notes of High Falconer* (1868), 2:486.

2. Two exceptions, based on archival sources, are J. W. Gruber, "Brixham Cave and the Antiquity of Man," in *Context and Meaning in Cultural Anthropology*, ed. M. Spiro (1965); and W. F. Bynum, "Charles Lyell's *Antiquity of Man* and Its Critics," *Journal of the History of Biology* 17 (1984): 153–87.

emerged from the efforts of a small core set of geologists. It was the product of careful fieldwork, but also of debates carried out in the privacy of meeting rooms and correspondence.

Brixham Cave and Paleontology, January–June 1858

J. W. Salter remarked in an 1858 letter to the editor of the *Geologist* that the progress of geology depended on detailed local research carried out by amateurs. Salter, then chief paleontologist to the Geological Survey, suggested that only amateur geologists had the time to devote to month- or season-long investigations of strata and fossils in areas near their homes but distant from London. If amateurs would undertake such studies and communicate the results to specialists working on the regions or eras in question, Salter claimed, "the more critical and obscure points which still remain *opprobria* in our British geology may be wonderfully cleared up"[3] The amateurs to whom Salter referred constituted one of the British geological community's greatest assets during the first six decades of the nineteenth century. Though their geological work frequently served a private, or local, scientific agenda, their willingness to share local observations made leading British geologists aware of information that might otherwise have gone unnoticed.

The investigation of the human antiquity question that began in 1858 depended on this symbiotic relationship between locally oriented amateur geologists and London-based career geologists. The discovery and excavation of Brixham Cave, which precipitated the investigation, demonstrate the significance of such cooperation. The cave's scientific value depended on a methodical, systematic excavation of its contents; only the coordinated actions of London and provincial geologists made such an excavation possible.

Brixham was, in the mid-nineteenth century, a pleasant but remote fishing village on the south coast of Devonshire. In January 1858, a local dyer named John Philp began to quarry Devonian-era limestone from a quarter-acre of land he owned on nearby Windmill Hill. On January 15, Philp's quarrying operation

3. J. W. Salter, letter, *Geologist* 1 (1858): 301.

opened a hole in what proved to be the roof of a previously unknown cave. Philp entered and explored the cave, locating one
of its original entrances and breaking through part of its stalagmite floor to discover a thick layer of earth studded with bones.
Philp was no geologist, but he recognized a potential tourist attraction. Within a few days of the discovery, he had cleared the
original entrance of debris, fitted it with a locked door, and set
up a makeshift glass case to display the bones he had uncovered.
By the time Philp opened for business, however, word of the new
cave had reached the neighboring town of Torquay, and the attention of an amateur geologist named William Pengelly.

The self-educated son of a Cornish fisherman, Pengelly
moved simultaneously in two scientific worlds: the local and the
national. He was a leading figure in the busy but relatively insular
intellectual life of Torquay. Settling there in 1836, he had become
a highly successful and respected day-school teacher and had
helped to reorganize the local mechanic's institute and found the
local natural history society. During the 1840s Pengelly acquired
a local reputation as a fossil collector, tutor in mathematics, and
lecturer on geology and astronomy.[4] By 1858 he was also an established member of the geological community. Elected to the
Geological Society of London in 1850, he began to attend its
meetings in spring 1855, and by 1856 he was also traveling to the
annual meetings of the British Association for the Advancement
of Science. Though not a career geologist, he could, by 1857,
count Andrew Ramsay, J. W. Salter, and other eminent scientists
among his acquaintances.[5]

Pengelly confined his geological activity almost exclusively to
the rocks of Devonshire and the neighboring county of Cornwall.
The four papers he published in the Geological Society's *Quarterly
Journal* and the dozen or more he published in the journals of
local societies before 1858 all concerned rocks from these two
counties. This geographically narrow focus may have begun as a
practical necessity, but by 1857 it had become a conscious choice.
Writing to his wife while on a trip to Ireland, Pengelly observed
that "we shall not do half the North of Ireland, but it cannot be

4. H. Pengelly, *A Memoir of William Pengelly* (1897), 16–36.
5. Ibid., 64–73.

helped, and I less regret it, as I have decided to devote myself to the geology of Devonshire." Two days later, he told her: "I have very much enjoyed my trip. . . . But why should I bother myself about the geology of this or any other distant district when there are so many unsolved problems connected with the pre-Adamic history of lovely Devonshire?"[6]

Pengelly's institutional allegiances reflected his intellectual interests. Despite his network of acquaintances in London, his closest ties were to his colleagues in Torquay. When he heard of the new cave on Windmill Hill, his first thoughts were of the Torquay Natural History Society. He approached Philp and found him "not disinclined to dispose of his Cavern, or rather the right of working it, to any person prepared to pay him well for it."[7] Pengelly had been part of the Torquay Natural History Society's 1846 excavations of Kent's Cavern, and in the new cave he doubtless saw an opportunity for further research along the same lines. The rest of the society's governing committee also expressed interest in an excavation of Brixham Cave, and Pengelly's description of their March 29 meeting makes it clear that they saw the cave primarily as a source of fossils representative of extinct local fauna. The society's leaders appointed a subcommittee and directed it to negotiate for a six-month license to dig in the cave and remove "whatever specimens, of any kind, they might find" for the Torquay Museum.[8] The outcome of these negotiations would ultimately force the Torquay Society to form a partnership with the Geological Society of London.

Pengelly's connections with the London-based geological community first became important to the investigation on April 16, when Dr. Hugh Falconer arrived in Torquay on a fossil-collecting trip. Falconer, a recently elected vice president of the Geological Society, had won the society's Wollaston Medal in

6. W. Pengelly to Lydia Pengelly, August 18, 1857, and August 20, 1857, ibid., 67–69.

7. William Pengelly, Brixham Cave Notebook 1, Pengelly Papers, Torquay. Pengelly's notes on the Brixham Cave excavations are written in thirteen paper-covered composition books, of which the third is missing. They were apparently composed after the excavation was complete—possibly in 1863—and include transcripts of letters and committee meetings. Unfortunately none of their pages is numbered (cited hereafter as Brixham Notebook).

8. Ibid.

1837 for his pioneering studies of the fossil mammals found in India's Sewalik Hills. By the time he retired from the Indian service in 1855, he had a reputation as a leading expert in vertebrate paleontology. In the following three years, he bolstered that reputation with a series of papers on the late Tertiary- and Quaternary-period fauna of Europe, particularly extinct species of elephant and rhinoceros.[9] Falconer had never met or even heard of Pengelly before he arrived in Torquay and saw Pengelly's name on an advertisement for an upcoming geological lecture. Nevertheless, Pengelly recalled, on "finding that I was a Fellow of the Geological Society [he] regarded me as a fellow labourer." Pengelly mentioned the new cave to Falconer, who visited it the next day with his colleague and long-time friend, Rev. Robert Everest. By the end of the week, Falconer, Everest, and Pengelly "were unanimous as to the probability that the cave was likely to be of value, and also that it must not be left in Philp's hands."[10]

Pengelly's dual intellectual life made him the pivotal figure in the early stages of the Brixham Cave exploration. He was a competent geologist, particularly familiar with the area around Torquay and nearby caves like Kent's Cavern, and so was well equipped to appreciate Brixham Cave's potential scientific importance. Because he lived nearby, he was able to inspect the cave soon after its discovery and to move quickly to prevent damage to its contents. At the same time, Pengelly's membership in the Geological Society meant that Falconer immediately accepted his knowledge of local geology and his assessment of the cave's potential importance. Falconer regarded Pengelly as a particularly skilled amateur—an ideal intermediary between the local and national scientific communities. Pengelly not only made data from Torquay available to the Geological Society, but served as a liaison between the London geologists and the citizens of Torquay. By the time Falconer and Everest left Torquay, they saw Pengelly as a fellow scientist united with them against Philp—a man with no scientific credentials.

Falconer had come to Devonshire to collect data for his ongo-

9. C. Murchison, "Biographical Sketch," in *Paleontological Memoirs*, xxv–xliv.
10. Pengelly, Brixham Notebook 1.

ing study of England's Pliocene and Post-Pliocene fauna. His correspondence makes it clear that, upon his return to London, he still saw Brixham Cave only as a source of paleontological information. Falconer began an April 20 letter to Lyell by discussing the extinct elephant and rhinoceros species he had found on his trip through southern England. His brief description of Brixham Cave and concluded with the hopeful statement that "from what I saw I expect a great harvest of bones." At the end of the letter, Falconer mentioned John MacEnery's detailed manuscript account of Kent's Cavern. He lamented the loss of the manuscript after MacEnery's death in 1836—not because of its intimations about human antiquity but because "nothing like it has yet been written" on cave paleontology. Falconer's May 9 letter to Joseph Prestwich opened with another discussion of elephant and rhinoceros fossils. It made no mention of Brixham Cave, focusing instead on the possibility of extending the distribution of "the new Rhinoceros" based on a newly explored series of deposits in Italy. Such an extension would, Falconer wrote, "give a vast lift in bringing out the distribution of the mammalian Fauna during the newer portion on the Pliocene series."[11]

The day after he wrote to Prestwich, Falconer submitted another letter to Searles Wood, secretary of the Geological Society. In it, he carried out his promise to Pengelly and the Torquay Natural History Society that he would ask the London society to provide funding for a thorough exploration of Brixham Cave. Philp had set his price for a six-month lease on the cave at one hundred pounds. The Torquay society had no hope of raising such a large sum, and so a methodical exploration of the new cave would only be possible if backed by money from elsewhere.[12] Falconer's letter, brought before the council of the Geological Society on May 14, thus began the transformation of Brixham Cave from a local curiosity into a nationally known geologic site.

Like his earlier letter to Lyell, Falconer's letter to the Geological Society focused entirely on the paleontological significance of

11. Falconer to Lyell, April 20, 1858, Lyell Papers, Edinburgh University Library (hereafter cited as EUL); Falconer to Prestwich, May 9, 1858, Falconer Papers, 318/2.

12. Pengelly, Brixham Notebook 1.

the cave. Falconer began by stating that virtually no work on bone-rich caves had been done in Britain since the publication of William Buckland's *Reliquiae Diluvianae* in 1823, and that the subject "has not advanced 'pari passu' with the progress in the investigation of the Upper Pliocene and Postpliocene deposits."[13] As a result, Falconer argued, serious misconceptions about the nature of British cave fauna had infiltrated the popular mind. The most damaging of these were the assumptions that all caves contain bones and loam deposited by the same agency at the same period, and that the fauna of one cave were essentially interchangeable with that of any other. These assumptions, he concluded, led paleontologists and fossil collectors to pay little attention to the stratigraphic position of cave bones. Worse, it caused provincial museums to lump bones from different caves together without any marks to indicate their provenance.

He went on to describe Brixham Cave, emphasizing that it was undisturbed and probably filled with fossil bones that represented a complete, intact extinct fauna. Falconer ended his description by asking "whether the case is not one deserving of a combined effort among geologists to organize operations for having it satisfactorily explored before mischief is done by untutored zeal and desultory work."[14] He explained that the importance of studying a virgin cave "had been forced upon" him during his recent attempts to establish the distribution of extinct elephants in Western Europe. To illustrate the point, he summarized the most important results from the cave-surveying tour that had brought him to Torquay and listed several questions that remained unresolved. A thorough investigation of "a well-filled virgin cave in England would materially aid in clearing up the mystery, either of the contemporaneity of the Pliocene mammalian fauna with the commencement of the Postpliocene fauna, or of the conditions and associations under which the former was replaced by the latter."[15]

13. Falconer to Secretary, Geological Society of London (GSL), May 10, 1858, in J. Prestwich, et al., "Report on the Exploration of Brixham Cave," *Philosophical Transactions of the Royal Society* 163 (1873): 472.

14. Falconer to GSL Secretary, in ibid., 473.

15. Falconer to GSL Secretary, in ibid., 474.

Two aspects of the letter deserve special emphasis here. First, Falconer repeatedly urged collective action by the Geological Society, emphasizing that Brixham Cave could solve problems that were of interest not just to a few geologists but to the entire geological community. Second, he argued explicitly that Brixham Cave was significant because its contents could help to resolve important paleontological problems. There is no evidence that Falconer initially regarded Brixham Cave as a test case for determining the age of the human race, or that he "tactfully avoided the question of the antiquity of man" in his letter.[16] Instead, he saw Brixham Cave as an opportunity to bring knowledge of cave paleontology up to the level of other stratigraphic and paleontological work on the Pliocene and Post-Pliocene deposits.

The governing council of the Geological Society was, like the British geological community as a whole, composed primarily of stratigraphers and paleontologists. It consisted of twenty-eight members, at least twenty-one of whom specialized in stratigraphy and paleontology.[17] Presented to such an audience, at a time when the British geologists' attention was already focused on the strata of the Post-Pliocene Epoch, Falconer's arguments proved compelling. He soon reported to Pengelly that the council had considered the proposal for a Brixham Cave excavation and "adopted it cordially," and that Sir Roderick Murchison had agreed, on behalf of the Geological Survey, to "back the proposal warmly." The council also voted tangible assistance. In the same meeting, it resolved to apply to the Royal Society for a hundred-pound grant and to form a "Cave Committee" to "look after the interests of Brixham Cave."[18]

The Geological Society's "Cave Committee" functioned, in effect, as a core set that had been given official corporate status. Its six original members all specialized in Pliocene and Post-Pliocene stratigraphy and paleontology, and all except Pengelly enjoyed

16. This assumption appears in G. Daniel, *150 Years of Archaeology* (1975), 57; and L. K. Clark, *Pioneers of Prehistory in England* (1961), quoted on 71.
17. The list of council members is taken from H. B. Woodward, *History of the Geological Society of London* (1907). The specialties are those listed in W. A. S. Sarjeant, *Geologists and the History of Geology* (1980).
18. Falconer to Pengelly, May 14, 1858, Brixham Notebook 1.

national reputations. Sir Charles Lyell, in addition to his status as a geological theorist and synthesizer, had defined the boundaries of the Pliocene and the other subdivisions of the Tertiary period. Richard Owen, professor of comparative anatomy at the Royal College of Surgeons, was the most respected vertebrate paleontologist of the day. Andrew Ramsay was a leading member of the Geological Survey and an expert on the Drift.[19] Joseph Prestwich had literally written the book on the Tertiary and Quaternary strata of England and France, and Hugh Falconer was Britain's leading authority on the fossil animals they contained. Pengelly, the only committee member who was not also a member of the council, knew more than any of the others about the contents and geological setting of Devonshire caves.

Other geologists joined the committee in succeeding months. Robert Godwin-Austen, whose work had unravelled the complex Recent strata of southwest England, added his stratigraphic expertise to Prestwich's. Samuel Beckles, a stratigrapher and vertebrate paleontologist who had done important work on Tertiary-period mammals, also signed on, as did Falconer's cave-hunting colleague Robert Everest.[20] John Phillips, professor of geology at Oxford, became a member of the committee largely because his status as president of the Geological Society required it. The committee's six original members remained its guiding force, however, forging the geological community's new consensus on human antiquity in their discussions and presenting it in their papers.

In planning the excavation of Brixham Cave, the members of the Geological Society's committee initially allotted themselves only a modest role. On May 16, two days after the committee

19. For Lyell, see M. J. S. Rudwick, "Charles Lyell's Dream of a Statistical Paleontology," *Paleontology* 21 (1978): 225–44; for Owen, see A. Desmond, *Archetypes and Ancestors* (1982), 19–23; for Ramsay, see Davies, *Earth in Decay,* 298–99, 333–35.

20. For Beckles's career, see A. Geikie, "Anniversary Address," *Proceedings of the Geological Society of London* 47 (1891): 54. There is no biographical information on Everest available, but in discussing their joint visit to Brixham, Falconer described him as "my friend . . . with whom I have been associated in similar objects, off and on, during thirty years." H. Falconer, "Primeval Man and His Contemporaries," in *Paleontological Memoirs,* ed. C. Murchison, 592.

had been formed, its members received word that the Royal Society had approved their request for a hundred-pound grant. Writing to Pengelly with the good news, Falconer suggested that the next step was to secure the "good will and cooperation" of the Torquay Natural History Society and to form a committee at Torquay to "direct and control" operations.[21] The operations to which Falconer referred were those related to the day-to-day work of excavating the cave. The Torquay committee was to ensure that the bones were removed from the cave and forwarded to London accompanied by details of their stratigraphic position and other important facts. The amateurs on the Torquay committee would thus expedite the collection of data that the specialists on the London committee would analyze.

The London geologists carefully assured their Torquay counterparts that the partnership between the two committees would be equal. Falconer thought it vital to satisfy the Torquay Natural History Society that "the Committee here . . . had no wish to cut in as interlopers, but to aid and cooperate with them in a very desirable scientific exploration."[22] The Geological Society resolution that formally created the Torquay committee empowered its members to act "in all subordinate matters . . . to the best of their discretion."[23] Most important, the London committee made it clear from the outset that they intended the Brixham Cave fossils to stay in Torquay. Even after the Royal Society laid claim to the fossils as a condition of its grant, Falconer reassured Pengelly that it probably wanted only a sample, and that the balance could go to the Torquay museum.[24] For the Torquay geologists, who saw the cave primarily as a source of fossils to illustrate local varieties of extinct animals, this was an important reassurance.

In the eyes of the London committee, directing and controlling the excavation of the cave did not oblige the Torquay committee to do any digging. Writing to Pengelly in mid-June, Prestwich recommended that local laborers be hired to remove the

21. Falconer to Pengelly, May 18, 1858, Brixham Notebook 1.
22. Ibid.
23. William Pengelly, Minutes of Brixham Cave Committee Meeting, May 25, 1858, Brixham Notebook 1.
24. Falconer to Pengelly, May 16, 1858, Brixham Notebook 1.

cave deposits, and that a "skilled collector" be sent from London to remove the fossils.[25] Henry Keeping, whom Prestwich hired to fill the latter position, belonged neither to the Geological Society nor to the geological community. He was a skilled laborer who worked for gentlemanly geologists like Prestwich and Pengelly on a contract basis, making a living from his ability to carefully and efficiently remove fossils from their matrix. In a letter written during the excavation, he told Pengelly that he did not intend to theorize about the cave, but to "endeavour to simply explain" to his employers those "passing events that may be daily taking place"—where particular bones had been found, for example.[26]

Keeping was the last man to join the team investigating Brixham Cave. By the time he arrived, the two committees had solved the last logistical problems connected with the excavation. Philp had signed a three-year lease on the cave, in which he granted the geologists the right to remove the sediment filling it and carry away any fossils they found there. Prestwich, as treasurer of the London committee, had disbursed twenty-five pounds from the Royal Society grant to cover the first lease payment, and another ten to cover expenses. Angela Burdett-Coutts, Pengelly's long-time friend, offered a further grant—gratefully accepted by the committee—of fifty pounds to cover the expenses of the excavation. Falconer, as chairman of the London committee, had reminded Pengelly of "the importance of the work and the necessity for great care and accuracy" and left him to plan the best way of excavating the cave. Keeping arrived from the Isle of Wight on July 14, and excavation of the cave began the following day.

Brixham Cave and Human Antiquity, July–October 1858

As originally conceived, the relationship between the Torquay and London committees mirrored, in miniature, the structure of the British geological community. The Torquay committee was composed of amateur geologists who were regarded as competent observers and fact collectors but as unsound theoreticians.

25. Prestwich to Pengelly, June 18, 1858, Brixham Notebook 2.
26. Keeping to Pengelly, August 23, 1858, Brixham Notebook 4.

Edward Vivian, a local banker who was the best-known member of the committee after Pengelly, is an excellent example. Vivian had published a number of descriptive papers on local geology, participated in the 1846 excavation of Kent's Cavern, and had partially reconstructed John MacEnery's lost Kent's Cavern manuscript. He was also, however, firmly committed to the geological reality of the biblical flood—a position that no leading geologist had defended for more than twenty years.[27] The Torquay committee was charged with overseeing the excavation, collecting the data that it produced, and forwarding them to London. The specialist and elite geologists on the London committee would play a relatively small role in the investigation until the time came to draw theoretical conclusions from the data.

In practice, the balance of power was different. Pengelly, who sat on both committees, had served as a liaison between the London and Torquay contingents since Falconer's visit in April. Because of this precedent, his acknowledged expertise in Devonshire geology and cave exploration, and his membership in the Geological Society, Pengelly became first among equals on the Torquay committee. Throughout the excavation, Falconer, Prestwich, and the other members of the London committee dealt with Pengelly to the virtual exclusion of everyone else in Torquay. From July 15, when the excavation began, until September 1, Pengelly was the only member of the London committee to set foot in the cave. On paper, the entire Torquay committee had been charged with overseeing the excavation and directing Keeping where and how to dig. In practice, only Pengelly maintained such autonomy.

Pengelly devised and implemented a revolutionary method of excavation, firmly based on geological principles, for the exploration of Brixham Cave. He insisted that the floor of the cave—a layer of limestone formed by calcium-rich water that dripped from the ceiling—be completely removed before the deposits below it were touched. As the floor was taken up, Keeping extracted the fossil bones that were imbedded in it, and Pengelly

27. E. Vivian, "On the Earliest Traces of Human Remains in Kent's Cavern," *BAAS Report* 26 (1956): 119–20.

measured the distance of each from the cave's entrance. Once the limestone floor of the cave had been cleared away, Pengelly and Keeping applied the same procedure to the deposits below. They removed one layer at a time, careful to note the locations of the bones it contained. For each bone removed from these lower deposits, Pengelly recorded not only its distance from the entrance, but also the stratum in which it had been found and its depth below the stratum's surface.[28]

In 1858, the standard method of scientific excavation was to sink vertical shafts through the deposits being studied. Archaeologists excavating round barrows, a type of burial mound, generally dug a single shaft in the middle of each, hoping to find a skeleton and grave goods at the bottom. Geologists and fossil collectors excavating in caverns used multiple shafts in different parts of the cave to determine where the richest deposits of fossils lay.[29] Compared to this "vertical" style of excavation, Pengelly's "horizontal" approach was slow, costly, and apparently inefficient, but it had an important advantage: it preserved the stratigraphic integrity of the deposits. The old style of excavation removed all the strata and fossils from the area of the shaft in a group, creating the possibility that fossils from different strata might accidently be mixed together. Pengelly's method removed only one stratum at a time, leaving no doubt that all the bones removed during a given phase of the excavation belonged to the same stratum.

Pengelly's documentation of the specimens removed from the cave was meticulous to a degree then unknown in geology or archaeology. It ensured that there could be no confusion, even long after the excavation, about where a particular specimen had been found. Each specimen was assigned a number and packed in a box labeled with the same number. Those found in the same place were given the same number and packed in the same box, so that each box corresponded to a single locality in the cave where specimens had been found. Finally, Pengelly kept a notebook in which he made an entry for each box, recording its num-

28. This description is from Pengelly's unpublished report on the excavation, quoted in J. Prestwich et al., "Excavation of Brixham Cave," 482.

29. Daniel, *150 Years*, 152–54.

ber, the specimens it contained, the date when they were found, and their location in the cave.[30]

Pengelly based his revolutionary system of excavation and documentation on the three-century-old principle of superposition. The principle, formulated by Nicholas Steno in the 1660s, stated that in any undisturbed group of strata the oldest are at the bottom and the youngest at the top. Steno's principle, which linked the age of strata to their physical relationship, was a cornerstone of stratigraphy and thus of mid-Victorian geology. It also implied that understanding local stratigraphy and the stratigraphic position of each species' fossils was the key to understanding how the fauna of a given region had changed over time. Pengelly knew from his experience in Kent's Cavern that the thin, interleaved deposits found in caves made it difficult to establish the stratigraphic position of cave fossils. He also knew, from the reception of earlier work on caves, that the traditional method of excavation rendered any generalization about the relative ages of cave animals suspect. He used new methods for the Brixham excavation in order to minimize such problems and later told Lyell that he would have resigned his superintendence if the new methods had not been adopted.[31]

The advantages of Pengelly's methods were immediately clear to the London committee. All of its members were experienced paleontologists and stratigraphers, and, like Pengelly, they knew from experience that the only way to make sense of the new cave was to document precisely the stratigraphic position of each fossil. Pengelly's one-layer-at-a-time style of excavation and careful recording of the depth of each new find promised such precision, and the committee members enthusiastically backed the innovations. The depth of this support and of the committee's faith in Pengelly became apparent in August when Edward Vivian challenged Pengelly's methods.

On August 20, Vivian visited the cave and asked Keeping to dig a vertical shaft through the deposits, which Keeping refused to do without Pengelly's approval. Vivian complained that the

30. Pengelly in J. Prestwich et al., "Excavation of Brixham Cave," 482. The records appear, in tabulated form, on pp. 499–516.
31. Pengelly to Lyell, April 11, 1863, Lyell Papers, EUL.

excavation was moving so slowly that the London committee's funds would be exhausted before the first layer had been dug out and said that he would take the matter up with Pengelly. Informed of the episode by Keeping, Pengelly wrote to Falconer for advice on August 22. Falconer's reply on behalf of the London committee arrived the following day—a tribute to the efficiency of the Victorian postal service. Falconer praised the excavation's progress, offered unconditional support for Pengelly's methods, and implied that alternative suggestions from members of the Torquay committee would be entertained but were unlikely to be adopted.[32] After seeing the letter, Vivian protested that he regretted the disagreement but that he had no intention "of being summarily cashiered" by Falconer—the same man who had assured him that the Torquay committee should take an active role in the excavation.[33] His anger was undoubtedly justified, but the die had been cast. The London committee had firmly and officially allied themselves with Pengelly, his methods, and his status as sole superintendent of the Brixham excavation.

Fortunately, the stratigraphy of Brixham Cave ensured that Pengelly's relatively slow method of excavation would still be productive (fig. 4). The limestone floor and the band of black mold that lay immediately below it in some parts of the cave were each less than a foot thick and limited in extent. In contrast, the third stratum was a bed of reddish-brown loam that extended throughout the cave at a thickness of two to four feet. In most parts of the cave, including all of the largest gallery, the loam bed either formed the floor or lay directly under the limestone crust. It contained, together with broken fragments of limestone and nodules of iron ore, most of the bones found in Brixham Cave.[34] Pengelly and Keeping's excavations thus encountered first a bone-studded limestone floor and, soon afterward, a bone-rich loam bed. The London committee, which saw Brixham Cave as a potential source of diverse paleontological information, received

32. Pengelly's description of these events, the only one available, comprises the entries for August 20–23 in Brixham Notebook 5. Also see Keeping to Pengelly, August 23, 1858, Brixham Notebook 5.

33. Vivian to Pengelly, August 26, 1858, Brixham Notebook 5.

34. Prestwich, et al., "Excavation of Brixham Cave," 485–88.

Figure 4. Cross-section of
the deposits in Brixham
Cave. The limestone
fragments and black mold
pictured here were not
present in all parts of the
cave (based on Prestwich,
et al., "Report on the
Excavation of Brixham
Cave," 1873).

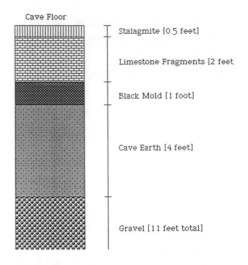

Cave Floor

Stalagmite [0.5 feet]

Limestone Fragments [2 feet]

Black Mold [1 foot]

Cave Earth [4 feet]

Gravel [11 feet total]

almost instant gratification. By the middle of August, after a
month's work, Pengelly reported to Falconer that 1,500 bones
had been removed from the cave. The excavation, as Falconer
told Lyell, was "going on famously."[35]

By mid-August, the loam bed that contained the bones had
also yielded something more startling: seven primitive tools made
of chipped flint. Pengelly and Keeping uncovered four during the
last days of July and three more in August, at depths of nine
inches to three and a half feet below the surface of the loam.
The tools were mingled indiscriminately with the bones of extinct
rhinoceroses, cave bears, and hyenas. In addition, all seven lay at
least a foot below the level at which the nearly complete skull
and antlers of a reindeer were imbedded in the limestone floor.
Twenty-nine more stone tools were uncovered in the cave before
the excavation ended the following year, but the original seven
were representative of the group.[36] Their stratigraphic position
was nearly indisputable proof of their age. Reindeer had not lived
in Britain since the end of the glacial period, in the middle of
the Post-Pliocene. The presence of reindeer remains in an unbro-

35. Falconer to Lyell, August 4, 1858, Lyell Papers, EUL. The report of 1,500
bones being found appears in a postscript dated August 24.

36. The dates and locations at which the tools were found are tabulated in
J. Prestwich et al., "Excavation of Brixham Cave," 494.

ken stalagmite floor made it likely that neither the floor nor the tools and bones beneath it had been disturbed since the Post-Pliocene.

The discovery of the stone tools changed the Brixham excavation, but the changes were subtle—matters of degree rather than of kind. No new members were added to the London committee after the tools were discovered, and no significant changes were made in the style of excavation. The excavation of Brixham Cave began as a geological investigation using geological methods, and it remained so even after the discovery of stone tools in the cave. Seen in the intellectual context of the late 1850s, the continuity makes perfect sense. The question of human antiquity turned on geological evidence—the stratigraphic relationships between human remains (either bones or implements) and the fossil remains of animals. Resolving the question was a matter of determining early humans' proper place in the fossil record. Since the 1820s, most geologists had treated humans as the climax of the Quaternary period's faunal succession and human remains as the end point of its fossil record. The fact that most geologists used the appearance of humans to mark the end of the Quaternary and of the geologic past highlighted the importance of early human remains in paleontology.

The discovery of stone tools did not change the geological character of the Brixham excavation, but it significantly increased the importance of the results. The nature of the boundary between Pliocene and Post-Pliocene faunas in Britain was an important question, but it was a technical problem of little interest to nonspecialists. The question whether humans had coexisted with extinct animals had implications not only for geology and other sciences but for religion as well—implications that were clear to scientists, clergy, and lay people alike. As a source of information about Pliocene paleontology, Brixham Cave had been valuable; as a site that might resolve the sixty-year-old human antiquity question, it was priceless.

The London committee exhibited little interest in the stone tools until a month or more after they were discovered, however. Pengelly was the only committee member immediately excited by them and by the implications of their stratigraphic position.

In early August, he wrote to Falconer—the only member of the London committee to take an active interest in the first stages of the investigation—and "particularly mentioned" the newly discovered tools. Falconer's response was uncharacteristically reserved. In a letter to Lyell begun on August 4 but not mailed until three weeks later, he summed up the report he had received from Pengelly with the comment that "although no striking results have been obtained, Pengelly thinks them encouraging." Several paragraphs later, his comment on an updated report from Pengelly was equally blasé: "This is a satisfactory report so far and quite as much as was to have been expected in the early stages." Falconer mentioned the flint implements only twice in the letter. Each time, he took less than a sentence to state simply that they had been found.[37]

The discovery of stone tools at Brixham Cave did not immediately change Falconer's ideas about the cave or its scientific importance.[38] He continued, throughout August, to regard the cave primarily as a source of paleontological information. When he stressed, in an August 18 letter to Pengelly, that they must "conduct the Brixham explorations in a careful and guarded manner, keeping an accurate record of the succession of the remains and their associations," his concern was not a direct result of the recent discovery of stone tools.[39] Instead, it was occasioned by Lyell's news of a primitive human skeleton that had been found in a stratigraphically uncomplicated deposit near Maastricht, Germany. Falconer saw the skeleton as compelling evidence that the stone tools from Brixham might corroborate but could not replace. He ended his August 18 letter to Pengelly as he had begun his May 10 letter to the Geological Society: insisting that the Brixham fossils be kept together at all costs in order to provide an unambiguous sample of Britain's Post-Pliocene fauna.

In early September, a month after he learned of the stone tools, Falconer suddenly developed an intense interest in them and in their value as evidence of human antiquity. His rapid shift from apathy to aggressive curiosity coincided with his second visit

37. Falconer to Lyell, August 4, 1858, Lyell Papers, EUL.
38. Cf. Gruber, "Brixham Cave," 389.
39. Falconer to Pengelly, August 18, 1858, Brixham Notebook 4.

to Brixham Cave, this time with Ramsay, which began on September 1. There is no record of what changed Falconer's mind about the importance of the stone tools, but the circumstantial evidence suggests two possibilities. Several years after the event, Falconer wrote that on his September visit to Brixham he identified the bones that had been removed from the cave. This firsthand experience may have convinced him that the bones found in the loam bed among the stone tools actually represented extinct species—a judgment Pengelly was not qualified to make. Simply *seeing* the evidence firsthand may also have spurred Falconer's interest. Fieldwork was central to British geologists' romantic image of themselves and their science, and data observed in the field thus possessed a cachet that equally reliable data gathered from printed sources or conversations did not. For Falconer, seeing the Brixham tools and bones in the field may have made them seem more tangible, more "real," and thus more important than they had seemed when Pengelly described them on paper.

Falconer's September visit to Brixham Cave did convince him that the chipped flints found there were genuine human artifacts, and that their presence among the bones of extinct animals was not accidental. He returned to London certain that the site could be important—perhaps even conclusive—in resolving the human antiquity question. Pengelly, who had made similar claims for the stone tools he had found in Kent's Cavern twelve years earlier, agreed with Falconer, as did Ramsay. The remainder of the London committee remained skeptical, however; they regarded the tools' status as human artifacts and the validity of the tool/bone associations as, at least, unproven. Falconer and Ramsay had gone to Brixham to see firsthand what progress the excavation had made, and the reception of the report resulting from their visit reflected the unresolved tension in the committee. Falconer, the report's principal author, submitted it to the London committee on September 9, 1858. Falconer's correspondence makes it clear, however, that the final version of the report was not released until more than a month had passed and a number of revisions had been made.[40] The committee-approved version

40. The final version of the report appears, dated September 9, as H. Falconer, et al., "Report of Progress," in J. Prestwich et al., "Excavation of Brixham

of the report, released in October, was a masterpiece of caution and careful wording. It discussed the chipped flints in terms that, while supporting Falconer's claims for the flints, were elastic enough to accommodate the skeptics' position as well.

The final report discussed the stone tools at greater length and in greater detail than any member of the committee had before. It noted that the tools had been found "in all parts of the cavern, mixed in the ocherous earth indiscriminately with remains of *Rhinoceros, Hyaena,* and other extinct forms," and that there was no evidence that the tools and bones were not contemporary. The final report also paid particular attention to the fact that one had been found thirty inches below the reindeer skull, labeling it a "result of great interest." Because the reindeer skull had been imbedded in the stalagmite floor of the cave, the report continued, it had almost certainly been one of the last additions to the cave deposits; since the stone tool lay well below it, humans had almost certainly lived in Britain before reindeer became extinct there. Finally, the official report stressed the care being taken at Brixham and suggested that "data will be arrived at for settling the disputed question" of whether the tools and bones had been deposited simultaneously. The report contained no mention of the paleontological questions the evacuation had initially been intended to solve.

Despite its extensive discussion of stone tools, the published version of Falconer's September 9 report was not a ringing endorsement of human antiquity. Every one of the theoretical conclusions it drew from the stone tools was modest or heavily qualified. It did not claim that the stone tools and bones of extinct animals had been deposited simultaneously—only that there was no evidence to the contrary. Given cave stratigraphy's potential to mislead, such a claim was not very compelling. The report described the chipped flints as objects "generally accepted at the present day as the early products of rude Keltic or pre-Keltic in-

Cave," 487. Falconer also gave September 9 as the date of the report in "Primeval Man," 593. Letters suggesting that the final, published version of the report was composed well after September 9 include Falconer to Pengelly, September 13, 1858, Brixham Notebook 6, and Falconer to Pengelly, October 11, 1858, Brixham Notebook 8.

dustry," but its discussion of them was headed "Human Industrial Remains(?)." In discussing the position of the reindeer skull, it concluded only that "the 'Reindeer' continued to be an inhabitant of Britain after the appearance of man in this island." Phrased in such a way, the statement was no declaration of human antiquity; it could just as readily support the proposition that the human race was young and the reindeer had died out in Britain later than most geologists had believed.

The relatively mild statements in the published report still troubled at least one member of the committee. Joseph Prestwich, intellectually conservative and as staunchly empiricist as the historical archaeologists, insisted that the report not be made public at the upcoming meeting of the British Association for the Advancement of Science in Leeds. Such a report, he complained to Falconer, "comes with a degree of might and authority which a short notice would not have. The statement that you make with regard to human industrial remains is one likely to give rise to so much controversy, and is one which you make so distinctly, that I do not like to see it embodied in a report which may be supposed to express the opinions of the several members of the Committee, and in which I see my name introduced." Prestwich went to say that Falconer made a good case, but that he hesitated to accept Falconer's conclusions until further work had been done. In particular, Prestwich said, he wanted to reserve judgment until he had *"worked* on the ground and looked at all the bearings" himself.[41] Prestwich's reply to Pengelly's request that the report be read at Leeds was similar: "A notice from you or Dr. Falconer . . . will be well, but any more formal report should in my opinion be kept to present to the public bodies who aid us."[42]

Prestwich's complaints were fruitless. On September 24, Ramsay read the cautiously optimistic progress report to a crowded meeting of the association's geological section, immedi-

41. Prestwich to Falconer, September 21, 1858, in G. Prestwich, *Life and Letters of Sir Joseph Prestwich* (1899), 117.

42. Prestwich to Pengelly, September 17, 1858, Brixham Notebook 6. Pengelly's letter to Prestwich is not extant, but it was clearly prompted by a September 14 letter from Falconer stating that he (Falconer) had no intention of presenting a paper, but that with Prestwich's approval the report could be read as an adjunct to the paper that Pengelly planned to deliver.

ately following a long paper by Pengelly on the geology of the cave. Pengelly's paper covered only those subjects that fell within his acknowledged field of expertise. In it he discussed the cave's structure and deposits in detail, but merely mentioned that two thousand bones and a handful of "what are supposed, and probably with much reason, to be flint implements" had been found. His paper, like the committee's report, was more a dispassionate summary of fact than an aggressively argued case for any theoretical position.[43] Richard Owen, also a member of the London committee, underscored this aspect of the two papers when he spoke on them, at some length, as president of the geological section. Owen's speech, Pengelly wrote later, was "chiefly to the effect that no discoveries had been made up to this time calculated to show that man is of higher antiquity than has commonly been supposed."[44] Together, the three speeches may have excited considerable interest, but they probably changed very few minds.

The first presentations of the Brixham Cave data to the geological community were not a turning point in the resolution of the human antiquity problem. They were important, however, because they established that the human antiquity question was, for the first time, the subject of active investigations by experienced geologists. The papers read at Leeds informed geologists outside the core set that the cave had yielded important data, but they endorsed no theoretical position that could not later be modified or abandoned. After the BAAS meeting, Falconer turned to the problem of convincing the skeptics on the London committee that the tools and bones found in Brixham Cave were conclusive evidence of human antiquity. To establish such a consensus, Falconer had to convince his fellow committee members of three things: first, that the chipped flints were human artifacts; second, that their stratigraphic relationship to the bones of extinct animals was not accidental; and third, that the relationship demonstrated the antiquity of the flints.

43. William Pengelly, "On a Recently Discovered Ossiferous Cavern at Brixham in Devonshire," Brixham Notebooks 6, 7. Only the title of the paper was published in the *BAAS Report;* Pengelly's manuscript copy of the text may well be the only one extant.
44. Pengelly, entry for September 24, Brixham Notebook 8.

Falconer made the first point by deploying the same evidence that had convinced him: the tools themselves. He wrote to Pengelly on October 5, asking him to send some of the "flint knives" to London—particularly the one that had been found directly beneath the reindeer skull. Three days later, he repeated the request, stressing that he needed the stone tools "to enable me to hold my own when assailed by skeptical people in the Committee."[45] Falconer evidently hoped that an opportunity to see, hold, and inspect the evidence firsthand would make as strong an impression on them as it had on him in early September. His hopes were fully justified. The tools arrived, and after examining them even Prestwich admitted that they could only be what the report suggested: human artifacts.[46] Falconer strengthened his case by appealing to experts in relevant specialties. He took several examples of the Brixham tools to the British Museum, where A. W. Franks, curator of British and medieval antiquities, "at once admitted their character i.e. that they bore as distinct proof of the hand of man as any pieces in the Br. Museum Collection."[47]

The London committee had, by October 12, reached a limited consensus. Its members agreed that the chipped flints found at Brixham were genuine human artifacts, though they were divided over the genuineness of the tool-bone associations and the actual age of the tools. The limited consensus was enough, however, to convince the committee that the cave was an important test case for the human antiquity question. This new realization wrought far greater changes in the Brixham excavation than did the initial discovery of the tools. During the summer, when the committee members believed that the excavation would yield important but routine paleontological data they had been content to cede control to Pengelly and (in principle) the Torquay committee. Once convinced, in mid-October, that the cave might provide a solution for the human antiquity question, they abandoned their laissez-faire style of management. Falconer began to

45. Falconer to Pengelly, October 5, 8, 1858, Brixham Notebook 8.
46. Falconer to Pengelly, October 11, 1858, Brixham Notebook 8; Falconer to Lyell, October 12, 1858, Lyell Papers, EUL.
47. Falconer to Pengelly, October 12, 1858, Brixham Notebook 8.

lecture Pengelly, in his letters, on proper excavation technique and the need for careful measurements. He sternly stressed the "very grave and serious importance" of the stone tools and their stratigraphic position, and reminded Pengelly—who hardly needed such reminders—that the committee had pledged to observe and record the contents of the cave "with severe fidelity." He also urged Pengelly to watch Keeping's work closely, lest the collector makes a careless error that might later cast doubt on the accuracy of all the data. Treating Pengelly as an amateur rather than an accomplished colleague, Falconer recommended that he "[go] over the details of position of every one of the flint knives in such a way that we could give an affidavit on them."[48]

Falconer's attitude toward Pengelly was part of his zealous attempt to ensure that the Brixham data would be unimpeachably accurate. Convinced that a solution to the human antiquity problem was at hand, he was determined not to have his case founder on an amateur's foolish or careless error. The same determination led the London committee to all but sever their ties with the Torquay committee after mid-October. The first blow to the Torquay group had been the London geologists' August defense of Pengelly's excavation methods and rebuff of Vivian's methods. The second came on October 11, in the form of a resolution by the London committee that it would be "inexpedient" to publish any data from the Brixham site until the cave had been more thoroughly explored. In principle, the resolution was designed to block the publication of any statements about the cave that the London committee was not willing to stand behind. In practice, it was a sharp reprimand aimed at Vivian for a September 29 article in the *Torquay Directory* that had blithely stated that the reindeer skull from Brixham was "of a very recent date," and that the cave's uppermost sediments had likely been deposited by the biblical Deluge. Vivian's article, like his earlier papers on Kent's Cavern, associated the cave evidence with what leading geologists regarded as outmoded, unscientific attempts to reconcile geology with scripture. This, the London committee quickly decided, was the last sort of publicity that the Brixham investigation needed.

48. Falconer to Pengelly, October 13, 1858, Brixham Notebook 8.

Within a week of the passage of the resolution against publications, the problem of what to do with the specimens from the cave arose for the second time. In May, when the excavation was first being organized, Falconer had assured the Torquay committee that they could have custody of most of the cave's contents. By mid-October, with the London committee convinced of Brixham's importance as a human antiquity site, Falconer had grown significantly less generous. He noted in a letter to Prestwich that Vivian "wants to get the specimens all into the Torquay Museum. They would be exposed there to be handled, and the tickets and labels might from careless usage drop off, and I do not think it would be advisable." Several days later, he offered Pengelly an even stronger opinion: "The Committee, after what has passed, will I am sure *not* allow the specimens to be lodged in the Torquay Museum on any terms. This is strictly *entre nous*. A single lable [sic] lost by careless handling, or a specimen gone astray, would be a fatal loss in such an important investigation." By November 3, the London committee had settled the issue, concluding that the site's newly recognized importance outweighed any earlier promises to the Torquay committee. It dismissed Vivian's plea on behalf of the Torquay Museum and resolved that, due to "the necessity of guarding the authenticity of specimens by no unnecessary displacement, it will be necessary to find a central point," not yet selected, to house the artifacts.[49]

The last major episode in the Brixham Cave excavation was played out in early November, when Prestwich went to Brixham

49. Falconer to Prestwich, October 18, 1858, Falconer Papers, 326/2; Falconer to Pengelly, October 23, 1858, Brixham Notebook 10; William Pengelly, "Report of the Brixham Cave Committee" Meeting, November 3, 1858, Brixham Notebook 10.

The issue of what to do with the bones and artifacts from Brixham Cave was not settled until after the final report on the excavation was written more than fifteen years later. Brixham was, by 1874, one among many sites that had yielded evidence of human antiquity, and the committee's concern for the safety of the finds had become less intense. The final report's authors recommended, in a letter to the Royal Society, that the flint tools be given to the Christy Museum, London's premier collection of prehistoric artifacts. They suggested that the British Museum be given its choice of the bones, and that the balance be sent to the Torquay Natural History Society in order to enrich its collection of local fossils.

J. Prestwich, et al. to Secretary, Royal Society of London, May 13, 1874, Royal Society Miscellaneous Correspondence, vol. 10, no. 108.

to "work on the ground and look at all the bearings." Accompanied by fellow stratigrapher and committee member Robert Godwin-Austen, he visited the cave on November 1. Like Falconer's visit two months earlier, Prestwich's sojourn marked the first time that he had seen the cave since the excavation began. The trip changed Prestwich's mind more quickly and completely than any number of detailed written descriptions. Discussing the visit in a letter to Falconer, he reported that "Austen is satisfied that the flint instruments occur with the bones. After my last visit I cannot deny it, but I am still not satisfied without seeking every other possible explanation than that of contemporaneous existence." He reminded Falconer, characteristically, that "we cannot be too cautious," accepting Falconer's claim that the tools actually lay among and below the remains of extinct animals while withholding judgment on their actual age.

By November 1, however, Falconer was no longer in England or in charge of the excavation. Obliged by his chronic rheumatism to spend the winter in a warmer, drier climate than London could offer, he had left for the Mediterranean. Though loath to leave while the Brixham excavation was still underway, he made a virtue of necessity and set out to collect data in southern Europe that would strengthen the case that he had built at Brixham Cave. Ironically, his first stop—at Abbeville, on the Somme River—changed the direction of the human antiquity investigation and eventually reduced Brixham Cave to a modest sideshow.

Hugh Falconer in Europe, November 1858–March 1859

Falconer left England in late October, bound for Sicily and traveling with his widowed niece, Grace Milne McCall. Their route, by careful design, took them through Montpellier, Nice, Lyons, and Abbeville and gave Falconer a chance to examine the sites and specimens central to the human antiquity debates of the 1830s and 1840s. The same scientific agenda influenced his choice of Sicily as a destination: there were numerous bone caves on the coast near Palermo. Falconer and McCall arrived at Boucher de Perthes's Abbeville home on the bitterly cold morning of November 1, and—though Falconer had only met de Perthes once, two years before—were received cordially. McCall remembered the scene vividly. "He [de Perthes] was just upon seventy, vigorous and active, not at all betraying his years. He looked a man carefully preserved: the thick, brown wig was unmistakably a wig, and there was a suspicion—nay, a certainty—of artificial coloring about his complexion. He showed us to his private study . . . which was literally crammed with curiosities. * * * The roomy old house was absolutely filled with relics and treasures of bygone days, with not a single habitable room in it, and must have been a dreary abode for any other than its owner." In time, de Perthes showed them his geological collections, including the chipped flints and elephant remains that he had dug out of the Somme Valley gravel beds in the 1840s. Nearly beside himself with excitement, he explained to Falconer in detail where each had been found.[50]

Falconer treated de Perthes as he had first treated Pengelly: as a trustworthy observer and an expert on local geology. He carefully examined de Perthes's cross-section of the gravel beds at Menchecourt, and identified as molars of the extinct mammoth several fossil teeth de Perthes had found there. Surveying a wide range of chipped flints from the same gravel beds, he dismissed many of them as accidents of nature, but noted that others were very similar to those found at Brixham Cave. He concluded that de Perthes, despite his absurd view of earth history and his over-active imagination, was right about one thing: the tools and extinct elephants' teeth from Menchecourt were contemporary. Having seen what he had come to see, he left Abbeville the same day.

Falconer said and did very little about Boucher de Perthes's specimens after he left Abbeville. He was convinced of their importance, but he was far more interested in evidence from caves than in evidence from "open air" sites like Menchecourt. This interest was consistent with Falconer's background: he had spent nearly three years studying the stratigraphy and fossil remains of Pliocene and Post-Pliocene caves but was not well acquainted with the Tertiary and Quaternary stratigraphy of France. Joseph

50. G. Prestwich, "Recollections of M. Boucher de Perthes," in *Essays Descriptive and Biographical* (1901), 83–85. Grace McCall married Joseph Prestwich in 1870.

Prestwich, in contrast, was Britain's leading authority on the French Tertiary and Quaternary strata, and Falconer wrote to him even before leaving Abbeville. In the letter, Falconer outlined what he had seen at Boucher de Perthes's house, and suggested that Prestwich would find a visit to Abbeville rewarding. He concluded, "You are the only English geologist I know who would go into the subject *con amore*. I am satisfied that English geologists are much behind the indications of the materials now in existence with respect to this walk of *Post-Glacial* geology, and you are the man to bring up the lee-way."[51]

Falconer had good reason to believe that the letter would pique Prestwich's interest. The two had discussed the human antiquity problem earlier in the fall, and Prestwich had cited Menchecourt as a potentially important site. Prestwich had spoken then of his "deferred intent" to investigate the evidence for human antiquity collected by others during the past ten years. He had also petitioned, unsuccessfully, to have some of the Brixham Cave grant diverted to fund the excavation of a promising "open-air" site at Brentford, in the neighboring county of Essex. Because the stratigraphy of cave deposits was often unclear, he believed that a careful excavation of a stratigraphically simple open-air site would do more to resolve the human antiquity problem than any number of cave excavations. Falconer's letter was also well timed. It reached Prestwich when he returned from his November 1 visit to Brixham and drew Prestwich's attention to an open-air human antiquity site in a region whose geology he knew well. Here, perhaps, was the "unmistakable corroboration" that he had wished for.[52] His reply was encouraging: "I am very glad you stopped at Abbeville, and am thereby fully confirmed to visit that locality at any early opportunity, and, as you suggest, to make the acquaintance of M. Boucher de Perthes."[53]

It would be easy to conclude, with the benefit of hindsight, that Falconer recognized de Perthes's specimens for the conclusive evidence they would eventually become. It is tempting to read his letter to the skeptical Prestwich as an exhortation to

51. Falconer to J. Prestwich, November 1, 1858, in G. Prestwich, *Life and Letters*, 119–20.

52. J. Prestwich to Falconer, September 21, 1858, in ibid., 117.

53. J. Prestwich to Falconer, February 4, 1859, in ibid., 121.

come, look, and be convinced.[54] The balance of Falconer's correspondence during this period will not support such interpretations, however. In letters to Pengelly and Lyell, Falconer discussed the specimens he had seen at Abbeville as though they
were important but not extraordinary. He described them in detail, but he devoted as much or more space to similar evidence
from the gravel beds near Lyons and the caves at Miallet. In fact,
by mid-November he was considerably less excited about de Perthes's stone tools than he was about a set from Miallet showing
the same form of chemical weathering as those found at Brixham.[55] The specimens from Menchecourt were, to Falconer, one
among many sets of data corroborating the results of the Brixham
excavation. Even his November 1 letter to Prestwich ended with
the statement, "What I have seen here gives me still greater im-
pulse to persevere in our Brixham exploration."

Falconer amassed a great deal of corroborative data on his
trip through southern France. He examined stone tools and human bones from caves near Montpellier, Nice, and Mentone and
learned—from a Mr. Charles Martin—of stone tools found with
remains of extinct animals in Scandinavia.[56] The crowning glory
of his field season, however, was his work on the Grotta di Maccagone near Palermo. Like Brixham Cave, the Grotta di Maccagone was a hitherto-unknown limestone cave filled with the remains of extinct animals; like Brixham Cave, it also contained
evidence of human occupation in the distant past. Broken stone
tools, charcoal, and bits of burnt clay were scattered through the
cave's stalagmite deposits, mingled with the remains of horses
and extinct hyenas. Falconer was impressed by the cave and its
contents, and by the "striking manner in which they confirmed
the results arrived at by the exploration of Brixham Cave."[57] He
announced the discovery in a March 20 letter to Lyell, an excerpt
from which was read to the Geological Society on May 4. In the

54. Examples of this position include Gruber, "Brixham Cave," 390–94; and
D. K. Grayson, *The Establishment of Human Antiquity* (1983), 184–85.
55. Falconer to Pengelly, November 14 and December 4, 1858, Brixham
Notebook 11; Falconer to Lyell, November 18 and December 8, 1858, Lyell Papers, EUL.
56. Falconer, "Primeval Man," 594; Falconer to Lyell, December 8, 1858,
Lyell Papers, EUL.
57. Falconer, "Primeval Man," 593–95.

intervening five weeks, however, the status of the human antiquity question had changed drastically.

Completing the Solution, April–August 1859

Joseph Prestwich left London for Abbeville on April 25, 1859,
following Falconer's advice that he should inspect Boucher de
Perthes's collections for himself. There can be little doubt that he
did so as a skeptic. His visit to Brixham in November had convinced him that stone tools and the bones of extinct animals were
found mingled in cave deposits, but not that this mingling proved
them to be contemporary. He would not admit the latter point,
he had told Falconer, until every other explanation of the Brixham data had been disproven. Since the stratigraphy of a cave
site like Brixham could be explained in many ways, it would be
impossible to conclusively eliminate those accounts in which the
tool-bone association was accidental. The tools and bones might,
for example, have been washed out of separate strata and
brought together in the cave by a flood. Prestwich's visit to Abbeville, like his abortive excavation at Brentford, was an attempt to
find a site that would yield a straightforward answer to the question: did humans and extinct mammals coexist?

Prestwich planned to go to Abbeville with a party of specialist
colleagues from the Geological Society—an informal "Abbeville
Committee" that could draw up a report on the subject. He could
not assemble such a party on short notice, however, and arrived
in Abbeville accompanied only by his friend John Evans.[58] Evans,
like Prestwich, was a businessman by trade and a scientist only in
his spare time. He was both a competent geologist and a respected
archaeologist—a nearly unique achievement for a scientist of his
generation—and a fellow of both the Geological Society and the
Society of Antiquaries. Flint implements lay far outside Evans's
archaeological specialty; his reputation rested on his excavation
of Roman-era sites and his studies of ancient British coins. As a
geologist he had worked with Prestwich on the age of the gravel
deposits that made up part of the Drift.[59] This experience, com-

58. J. Prestwich, "On the Occurrence of Flint-implements," *Philosophical
Transactions of the Royal Society* 150 (1860): 281–82.
59. Joan Evans, *Time and Chance* (1943), 83–97.

bined with his long-standing friendship with Prestwich, resulted in his invitation to Abbeville.

Prestwich left England on April 25, and proceeded to Abbeville, where Evans joined him the next day. Evans's journal shows that he left for Abbeville well aware of the recent discoveries in Devonshire and Sicily and of the potential implications of the phenomena he and Prestwich were to investigate. "Think of their finding flint axes and arrowheads at Abbeville in conjunction with the bones of Elephants and Rhinoceroses 40 ft. below the surface in a bed of drift. * * * I can hardly believe it. It will make my ancient Britons quite modern if man is carried back to the days when Elephants, Rhinoceroses, Hippopotamuses and Tigers were also inhabitants of the country."[60] They visited Boucher de Perthes's home, as Falconer had, and were presumably shown the same array of stone tools, bones, and cross-sections. Prestwich, though presumably interested, was not satisfied. Consistent with both his own cautious empiricism and the geological community's emphasis on evidence observed in the field, he wished to see a stone tool still imbedded in the gravel among mammalian remains. He and de Perthes searched for such a tool at Menchecourt, Moulin-Quignon, and several other sites near Abbeville, but without success. Prestwich moved on to the gravel pits at St. Acheul, near Amiens, where "long fresh faces of gravel afford, together with the digging for gravel in daily operation, ready and convenient sections for observation," but again he found nothing. Prestwich departed, leaving instructions with the Amiens workmen to notify him of any new discoveries. Almost as soon as he had returned to Abbeville, a telegram arrived from Amiens, stating that a stone tool had been uncovered and left in situ for him to examine.

Taking Evans with him this time, Prestwich returned to Amiens. There they examined the newly discovered tool, took detailed notes, and arranged for photographs of the gravel face where it had been found. This single observation was enough to validate de Perthes's claims of many similar discoveries elsewhere along the Somme. Having arrived on April 25 and 26, Prestwich and Evans returned to England on April 29.[61]

60. Quoted in ibid., at 100.
61. Prestwich, "Flint-implements," 282–93.

Prestwich returned from Abbeville convinced that humans
had coexisted with extinct mammals. As Falconer had after visiting Brixham Cave eight months before, he immediately made
plans to announce the crucial evidence to his scientific colleagues. He began on May 5, inviting geologists and archaeologists to his house for sherry and a first look at the implements he
had collected in France.[62] Working quickly, he submitted a paper
on the Somme Valley evidence to the Royal Society on May 19
and read it at the society's May 26 meeting. The paper began with
a brief resumé of earlier work on the human antiquity question,
emphasizing the discoveries at Brixham and their promising but
inconclusive nature.[63] It went on to describe the sites he had investigated on his recent trip to France: Menchecourt, Moulin-
Quignon, St. Roch, and St. Acheul. Prestwich's descriptions of
the sites were clearly the work of a geologist treating a geological
problem; like Falconer's report on Brixham Cave, Prestwich's paper discussed the stratigraphy and paleontology of each site in
detail. It paid particular attention to the species of fossilized fresh-
water and marine molluscs found at each site—data that Lyell
had used to subdivide the Tertiary period into epochs—and that
remained important in attempts to decipher Tertiary and Quaternary stratigraphy.[64]

The same geological approach was apparent in the heart of
Prestwich's paper, a detailed discussion of "the nature and value
of the evidence." The discussion was clearly designed to refute,
as a group, the most venerable objections to the idea that humans
and extinct mammalia coexisted. Prestwich first asked whether
the apparent stone tools might simply be the result of natural,
accidental fracture patterns. He then asked whether they could
be "of recent manufacture"—that is, the product of some primitive but comparatively recent tribe. Finally, he asked whether the
stone tools could have been derived from a younger bed, or the
fossil bones from an older bed, than that in which they are now
mingled.[65] In each case, Prestwich answered "no." In each case,

62. Evans, *Time and Chance,* 122.
63. Prestwich, "Flint-implements," 277–82.
64. Ibid., 282–94; Rudwick, "Statistical Paleontology."
65. Prestwich, "Flint-implements," 294.

he supported his answer with arguments based on detailed geological knowledge and careful observation of the tools' geological setting.

Prestwich attacked the idea that the tools could have been
accidents of nature by reminding his audience that "the argument does not rest upon the evidence of skill, but upon the evidence of design." He went on to describe the characteristic fracture pattern of the flint from which the tools were fashioned.
Because it has "no tendency to break in one direction rather than
in another," a piece of flint chipped by random collisions should
be irregular, with no pattern to its fractured surfaces. Nearly all
of the flints found at Abbeville and Amiens had, in contrast, one
of three distinct, complex forms (fig. 5). This regularity of structure, he told his audience, "surely implies design, the application
of forethought, and an intelligent purpose." Prestwich concluded
that not all of the flints in de Perthes's collection showed such
clear evidence of human manufacture, but that it "suffices for our
purpose, in treating of the geological question, that we have one
set of implements showing art and design."[66]

Prestwich then turned to the question of whether the tools
had been manufactured recently. The surfaces of flints, he
pointed out, take on characteristic chemical stains from the sediment in which the flints are imbedded; in some cases they become white, in others yellow, and in still others brown. These
stains were, in fact, a standard geological tool for precisely match-
ing flint nodules with the beds from which they were derived.
The flints in de Perthes's collection, Prestwich told his audience,
"present a colouring in perfect accordance with the lithological
character of the beds of which they formed a part." He argued
that the chemical stains, like the small grains of sediment cemented to the flints and the calcareous film that covered even
their fractured surfaces, were clear evidence of the tools' great
age. Such phenomena would only occur if the flints had lain for
"considerable time" in the beds from which they had been extracted.

The third question—whether the association between tools

66. Ibid., 295–96.

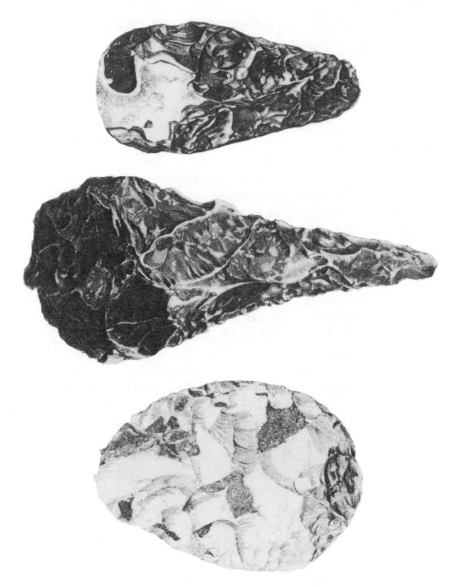

Figure 5. Chipped-stone tools from the Somme Valley, showing the three characteristic forms that Evans described (from Evans, "On the Occurrence of Flint Implements," 1860).

and fossilized bones proved their contemporaneity—was the most important. It was also the most disputed. Almost all of the objections to the human antiquity theories of the 1830s and 1840s stemmed from the idea that the tools and bones now found together had been deposited in widely separated beds. Prestwich himself had finally accepted the coexistence of humans and extinct animals only after inspecting the sites along the Somme Valley. He dismissed the possibility that the tools had been inserted in the gravel beds long after the beds' deposition by invoking his own geological expertise. "To anybody accustomed to the examination of drift deposits," he told his audience, "there is little difficulty in distinguishing between the fresh and uniform appearance of undisturbed beds, and the mixed and confused [appearance] of made ground." [67] The stone tools of Abbeville and Amiens, he continued, came from beds that were clearly undisturbed. Moreover, the beds above them were also intact—the thin, laminated layers of sand and the delicate shells they contained had not been disrupted (fig. 6). Finally, Prestwich noted, all the tools found in the Somme gravel beds lay with their flat surfaces parallel to the surface of the beds. Had they been thrust in from above—or fallen in through cracks in the ground—they would, instead, have been perpendicular to the beds. [68]

Having established that the tools were contemporary with the gravel beds, Prestwich turned his attention to the bones, giving special attention to the possibility that they had come from some lower (and thus older) bed that had been dissolved or torn up during the deposition of the gravel. To refute this objection, Prestwich presented six geological arguments. First, he told his audience, the fragments of bone from the gravel bed showed fewer signs of wear and rounding than did the tools; had they been washed out of an older bed and redeposited, they should have been more worn than the tools. Second, no chemical stains or attached bits of sediment indicated that the bones had once lain in a different bed. Third, no other beds in the district could have served as the source of the bones. Fourth, the bones lay alongside delicate fossil shells that were unquestionably of the

67. Ibid., 299.
68. Ibid., 299–300.

Figure 6. Cross-section of the deposits at a typical human antiquity site in the Somme Valley gravel pits. In surrounding areas that had not been quarried, an eight-foot layer of earth covered the deposits shown. The details shown here are those of the pit near St. Acheul where Prestwich and Evans first saw a stone tool in situ (based on Prestwich, "On the Occurrence of Flint-implements," 1860).

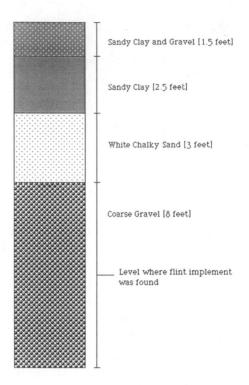

Sandy Clay and Gravel [1.5 feet]

Sandy Clay [2.5 feet]

White Chalky Sand [3 feet]

Coarse Gravel [8 feet]

Level where flint implement was found

same age and that could not have survived being washed out and redeposited. Fifth, de Perthes had found the leg bones of an elephant and the complete skeleton of a rhinoceros in the gravel—an impossible occurrence if the bones had washed out of another bed. Finally, the Somme gravels contained only extinct mammals normally found in the Drift; had they been washed out of an older bed, then species peculiar to the Tertiary should have been found with them.[69]

The balance of Prestwich's paper discussed the site of Hoxne, Suffolk, where John Frere had discovered flint implements in 1797. The site came to Evans's attention shortly after he and Prestwich returned from France, and the two had visited it in order to gather corroborative evidence. Prestwich had been eager to visit Hoxne before delivering his paper to the Royal Society,

69. Ibid., 300–302.

and he devoted considerable space to what he saw there. On the basis of the local stratigraphy and the presence of certain species of fossil molluscs, he concluded that Hoxne was analogous to the sites he had seen in the Somme Valley. The Hoxne site confirmed the conclusions that Prestwich had formed at Amiens: that the chipped flints were human artifacts, that they were found in undisturbed strata with the bones of extinct mammals, and that the toolmakers had once coexisted with the extinct beasts. Prestwich's work at Hoxne also assured him that the tool-bearing beds were younger than the extensive glacial deposit known as the Boulder Clay. In other words, the manufacturers of the Somme Valley and Hoxne stone tools lived after the end of the glacial period.[70]

Prestwich's May 26, 1859, paper was a long, detailed argument. It marked his acceptance of the idea that humans and extinct animals had coexisted and was designed to convince his Royal Society audience that they, too, should accept the idea. Prestwich did not stop there, however, any more than Falconer had rested after his report on Brixham Cave had been read to the British Association for the Advancement of Science the previous September. Like Falconer, he set out to bring his colleagues face to face with the evidence that had convinced him. On May 29, he returned to Amiens and Abbeville with Godwin-Austen and two other geologists—John W. Flower and Robert W. Mylne. The climax of this visit came at St. Acheul, when Flower discovered a large stone tool in the gravel bed, twenty-two feet below the original ground surface. Prestwich regarded the discovery as "an important and conclusive fact," and particularly stressed the fact that Flower has removed the tool from the gravel *"with his own hands."*[71] Flower, Mylne, and Godwin-Austen thus became, like Prestwich and Evens, eyewitnesses to the presence of human remains in the Drift gravels.

Flowers discussed his find in a brief, descriptive paper that he read before the Geological Society on June 22. At the same meet-

70. Ibid., 302–8.

71. Evans, *Time and Chance,* 122–23. Prestwich, "Flint Implements," 294n. The published version of Prestwich's May 26 paper contains several notes that were inserted before publication in order to discuss later developments.

ing, Falconer gave a detailed account of his discoveries at the Grotta di Maccagone, and displayed samples of the fossils he had found there. Prestwich, who had taken over nominal control of the Brixham excavation when Falconer left for Sicily, presented a summary of its principal results. The three papers were scheduled together by design, and the society postponed its summer recess for a week so that an "extraordinary meeting" could be held to accommodate them.[72] This may well have been the work of Prestwich and Falconer, both of whom sat on the society's council. The papers were the first formal discussion of human antiquity brought before the Geological Society in more than a decade, and they formed a smoothly interlocking argument in favor of the idea that humans and Quaternary mammals had coexisted. Each paper concerned a site that had been investigated in detail, and each was presented by a geologist who had examined the evidence firsthand. Writing to his wife from London on the day of the meeting, Pengelly remarked that "flints are certainly to the fore."[73]

Flints were also coming to the fore outside the geological community. On June 2, a week after Prestwich delivered his paper to the Royal Society, Evans read his own account of the Somme Valley discoveries to the Society of Antiquaries. Evans tailored his paper for his archaeological audience—discussing the different types of stone tools found and speculating on their possible uses—but it remained heavily geological. He began by arguing that archaeology and geology were kindred sciences and that geological discoveries could have important archaeological implications. He then turned to the discoveries in the Somme Valley and to those at Hoxne, as Prestwich had in his paper. In each case, Evans gave a detailed outline of the stratigraphy and characteristic fossils of the area. Using many of the same arguments Prestwich had used, Evans showed that the facts he had presented could best be explained by assuming the coexistence of humans and extinct animals.[74]

Evans drew on both his geological and archaeological experi-

72. W. Pengelly to Lydia Pengelly, 22 June 1859, in H. Pengelly, *Memoir*, 87.
73. Ibid.
74. John Evans, "On the Occurrence of Flint Implements," *Archaeologia*, 38 (1860): 280–307.

ence in his June 2 paper. He was addressing archaeological concerns—the nature of the stone implements and the culture to which they belonged—but he was doing so as a geologist and using far more geological data than any contemporary archaeologist would have. This fusion of geological means and archaeological ends formed the foundation of a new style of archaeology, of which Evans became the patriarch. The "geological archaeology" introduced in Evans's paper dominated British studies of prehistory from the mid-1860s until the end of the century.

The papers of Prestwich, Flower, Falconer, and Evans presented the geological community with a firm consensus on the question of whether humans and extinct animals had coexisted. Equally important, those discussing the French sites supplied evidence for human antiquity that was more convincing than any that had emerged before. The evidence from Brixham Cave had been sufficient to convince Falconer, Pengelly, and others whose specialties bore directly on the human antiquity question; it had even erased all but the last of Prestwich's doubts. Hoxne and the Somme Valley provided evidence that nonspecialists and even nongeologists could appreciate. It was on those sites that the members of the human antiquity core set built their new consensus.

The British Association Meetings, September 1859

The annual meeting of the British Association for the Advancement of Science was usually the largest scientific gathering of the year in Britain. Most of the geological community's leading members, along with scores from its rank and file, attended the meetings of the association's geological section. The 1859 meeting in Aberdeen was large even by British Association standards, in part because Prince Albert had agreed to serve as president. It drew approximately 2,700 people, including virtually all of the geological elite: Charles Lyell, Roderick Murchison, Adam Sedgwick, Richard Owen, John Phillips, Andrew Ramsay, and a transplanted American, Henry Darwin Rogers, among them.[75] The Aberdeen meeting was thus an ideal forum in which to present the

75. Pengelly described the meeting and its participants in a series of letters to his wife dated September 16–21, 1859. Some of the letters are reprinted in H. Pengelly, *Memoir,* 89–91.

new consensus on the coexistence of humans and extinct mammals to the widest possible scientific audience. Lyell, president of the geological section, did so in his opening address.

Lyell came late to the investigation of human antiquity. Despite his membership on the Brixham Cave Committee from the moment it was formed, he took no part in directing the excavation and did not visit the cave until well after the work was done. Falconer, in a wistful letter dated December 8, 1858, complained that "I wish you would . . . be done with these weary volcanoes and return to your old love [Tertiary and Quaternary strata]. I think it hopeless to excite your interest at present in the possible connection of humans with extinct animals in the glacial period." Lyell's interest in human antiquity had only begun to reemerge by spring 1859. Before discussing his work at Palermo, Falconer told Lyell that "first you must go to school and learn a little" about "a class of geological remains which I suspect you have never yet studied and which . . . you have been in the habit of treating with a spice of unphilosophical scorn." Two weeks later, he recommended that Lyell visit the British Museum, so that he might learn to distinguish between naturally and artificially chipped flints.[76]

Despite his lack of extensive experience with the evidence of human antiquity, Lyell was the ideal person to present the new consensus. First, he had gone to France in July to examine the Somme Valley sites for himself and thus could speak from firsthand experience about the crucial evidence. He was, in fact, the only member of the Geological Society's cave committee who had visited the French sites and was also at the British Association meeting.[77] More important, Lyell was acknowledged by geologists and nongeologists alike to be a member of the elite group that

76. Falconer to Lyell, December 8, 1858, March 20, 1859, and April 6, 1859, Lyell Papers, EUL.

77. I have been unable to establish this with complete certainty, but it seems highly probable. Pengelly wrote to his wife regularly during British Association for the Advancement of Science meetings and made a point of telling her what colleagues he had met. The letters that he wrote during the Aberdeen meeting are highly detailed, but in them he never mentions seeing Prestwich, Godwin-Austen, or Falconer. Pengelly knew all three men, and Lydia Pengelly had met Falconer and Prestwich during their visits to Torquay. It seems highly unlikely that Pengelly would have seen them and not mentioned it in his letters.

could pass judgment on the broadest issues of geological theory. Even his detractors admitted that he was among the leading geologists of the age. His opinion on the question of human antiquity, backed by his newly acquired firsthand knowledge of the evidence, carried more weight than that of any other geologist possibly could.

Lyell's address—delivered on September 16, the first day of the meeting—was brief, concise, and tightly argued. Ostensibly a summary of important geological problems and recently proposed solutions to them, it was primarily a discussion of human antiquity. Lyell began by noting the many earlier attempts to establish the coexistence of humans and extinct animals and the revival of interest in the subject during the past few years. He devoted a few sentences to Brixham Cave, arguing that the results of its excavation "must, I think, have prepared you to admit that scepticism in regard to the cave-evidence in favour of the antiquity of man had previously been pushed to an extreme." Lyell quickly moved on, however, to the Somme Valley evidence for human antiquity, referring interested listeners to the abstract of Prestwich's May 26 paper for "a clear statement of the facts." He declared himself "fully prepared to corroborate" Prestwich's conclusions and noted that he himself had obtained an abundance of flint implements during his own visit to Amiens and Abbeville.[78]

Lyell concluded that, although the shells found along with them were of living species, "I believe the antiquity of the Abbeville and Amiens flint instruments to be great indeed if compared to the times of history or tradition." The geology of the Somme River valley, and the presence of now-extinct mammals in the tool-bearing beds implied, he told his audience, "a vast lapse of ages, separating the era in which the fossil implements were framed and that of the invasion of Gaul by the Romans."[79] Lyell's final sentence suggests that he, like Evans, was aware of the archaeological implications of his geological ideas. His choice of words separated the earliest Europeans from the earliest Euro-

78. C. Lyell, "On the Occurrence of Works of Human Art," *BAAS Report* 29 (1859): 93–94.
79. Lyell, "Works of Human Art," 95.

pean legends and traditions not by generations, but by untold thousands of years.

Lyell stood before the British Association little more than a year after the first flint tools were discovered in Brixham Cave and less than six months after Prestwich and Evans had visited Abbeville. His address announced, publicly, the consensus on human remains in the Drift constructed privately by a small group of expert geologists. It summarized, for the broadest possible scientific audience, what Prestwich, Falconer, Flower, and Evans had announced to the specialist societies in May and June. Together, those papers marked the climax of an intense investigation of—and debate about—human antiquity, but they ended neither the one nor the other. Instead, they expanded the scope of the investigation and made the debate public for the first time.

FIVE

Defending and Extending
the Consensus

THE PAPERS presented by Prestwich, Evans, Falconer, Flower, and Lyell in spring and summer 1859 made a strong case for human antiquity. More compelling than the arguments presented in the 1820s, 1830s, and 1840s, the new case quickly convinced a number of leading scientists. Its impact on the geological community was particularly pronounced. There, the new argument, constructed by men whose special expertise was acknowledged by all British geologists, was accepted almost as soon as it was presented. After June 1859, no British geologist publicly disputed the idea that humans and extinct animals had once coexisted. Historians have assumed, perhaps for that reason, that the 1859 papers were the last significant event in the process of establishing human antiquity.

The core set's actions and correspondence during the years between 1859 and 1863 demonstrate that they saw things very differently. Prestwich, Falconer, and their colleagues neither regarded their case for human antiquity as complete nor dismissed objections to it as insignificant. They embarked, in summer 1859, on an extensive series of investigations designed to gather more data on humankind's place in the geologic record. They also took pains to defend the quality of the evidence they had already collected, responding decisively to a series of attempts to discredit it. Not content to rest on the success of their 1859 papers, they displayed the evidence from Brixham Cave and the Somme Valley to members of various scientific societies in an ultimately successful effort to build support for their position.

This process of defending and extending the consensus built in 1859 came to a climax in 1863. The geologists' data-gathering campaign was so successful that, in April of that year, they were able to reject an apparently crucial piece of new evidence without weakening their case. Charles Lyell's *Geological Evidences of the Antiquity of Man*, published in 1863, quickly became one of the most popular scientific works of the decade. It presented the new case for human antiquity and displayed—for educated nonscientists as well as scientists—the wide range of newly gathered evidence for human antiquity. *Antiquity of Man* also marked the beginning of an era in which research on the earliest humans would be a cooperative endeavor involving geology, archaeology, and anthropology.

Defending the Data, July–December 1859

The case for human antiquity presented by Prestwich and his colleagues in spring and summer 1859 proved convincing because of the evidence on which it rested. The carefully structured arguments Prestwich outlined in his May 26 paper were important but subsidiary. In the resolutely empiricist world of Victorian science it was facts—observations made in the field—that made the paper compelling. The observations on which the case for human antiquity rested fell into three categories: those demonstrating that the chipped flints were human artifacts, those establishing the flints' association with the bones of extinct animals, and those showing that the associations were genuine. The members of the core set knew that doubt cast on any of the three types of observations would seriously damage the credibility of the case for human antiquity. Between June and December, 1859, nongeologists publicly criticized each of the three types.

Charles Babbage launched the first attack in a paper read to the Royal Society on May 26, 1859. Babbage, best known as a mathematician, was master of many subjects, but geology was not among them. He tailored his paper accordingly, saying little about specific associations of stone tools and extinct animals and objecting to recent claims for human antiquity solely on methodological grounds. The paper was inspired by Falconer's work on the Grotta di Maccagone, and Babbage was not aware of Prestwich and Evans's visit to France in April. It asserted that, while human artifacts may have been found with the remains of extinct animals, "it is certainly premature to assign [a] great antiquity to our race, as long as the occurrence of such mixtures can be explained by known causes admitted to be still in operation."[1] Babbage described several hypothetical cases in which "known causes" could bury human and mammalian remains from different periods in a single, apparently undisturbed, stratum. He did not claim that his hypothetical explanations were true, only that they were "open to less objection" than theories of human antiquity.

Babbage's paper, concerned primarily with cave evidence, made an important philosophical point. It was, in many ways, a formal restatement of the reservations about Brixham Cave that Prestwich had expressed to Falconer in September 1858. Its implicit appeals to Occam's Razor and to Lyell's insistence that all geological change be explained by causes now in operation put it on solid philosophical foundations. Nevertheless, it drew no specific published response from the geologists whose work it criticized.

There were two reasons for this nonreaction. First, Babbage's attempts to explain the tool-bone associations by causes now in operation were unconvincing. To account for Falconer's discoveries in Sicily, Babbage postulated the existence of a second cave—above, behind, and connected with the Grotta di Maccagone—that could serve as a source of fossil bones. He also postulated, on the basis of slender or absent evidence, a complex series of geological changes. His explanations for the tool-bone associations observed by de Perthes in the Somme Valley required similarly complex assumptions, and most scientists probably found it easier to accept human antiquity than to grant them. Second, Babbage read his paper to the Royal Society immediately after Prestwich's discussion of his own recent visit to France. Prestwich's paper anticipated, and refuted, all of Babbage's alternative explanations for the tool-bone associations in the Somme Valley before Babbage could present them. Equally important, Prestwich's new discoveries in France meant that the new case for

1. C. Babbage, "Observations on . . . the Remains of Human Art," *Proceedings of the Royal Society* 10 (1859): 59.

human antiquity no longer depended solely on cave evidence. Babbage's paper was obsolete before it was read.

Thomas Wright, one of Britain's best-known archaeologists, launched the second attack in a letter to the widely read weekly, *Athenaeum*. Wright's letter, which appeared on June 18, 1859, was a direct response to the papers Prestwich and Evans had read to the Royal Society and the Society of Antiquaries. It castigated both Prestwich and Evans for "coming rather hastily to conclusions" on the human antiquity question. There was strong evidence, Wright argued, that the chipped flints found in the Somme gravel beds were not human artifacts but accidents of nature. Having examined the specimens that Prestwich displayed to the Royal Society, Wright noted that they "presented forms not common among the flint implements ascribed usually to the Celtic period, with a total absence of . . . what we call finish." Prestwich's "tools," he concluded, were simply stones that had been shaped by countless random collisions with other stones on the river bed.[2]

Wright's objection seems merely obstructionist today, but to the archaeologist of 1859 it was legitimate. Though very few British archaeologists studied pre-Roman artifacts extensively in the 1850s, all were familiar with the existence of polished stone axe and hammer heads. These artifacts, called *celts* after the people believed to have made them, turned up regularly in barrows, chambered tombs, and other ancient grave sites. The three national archaeological journals commonly featured descriptions and illustrations of new ones. When British archaeologists discussed "stone tools" during the 1850s, they referred to the carefully worked and polished celts (fig. 7). Chipped-stone tools—found primarily in bone caves rather than in the barrows and tombs that pre-Roman archaeology studied—lay outside most archaeologists' experience. An inexperienced observer, used to dealing with smoothly finished stone tools, might well conclude that what the geologists called chipped-stone tools were naturally broken rock (fig. 8).

John Evans answered Wright—and appealed implicitly to others with similar doubts—in the June 25 issue of the *Athe-*

2. T. Wright, "Flint Implements in the Drift," *Athenaeum*, June 18, 1859, 809.

Figure 7. A typical chipped-stone tool (from Evans, "On the Occurrence of Flint Implements," 1860).

naeum. He conceded that the Abbeville flints were different from the polished stone tools of "later periods," but argued that they "bear upon them quite as evident marks of design in their formation." These marks—sharp, even edges and uniform shapes— "could not possibly" have appeared accidentally. That much of Evans's reply restated themes from his June 2 Society of Antiquaries paper. The bulk of his letter, however, appealed directly to the phenomena, as Falconer had done in swaying the Brixham Cave Committee the previous fall. Evans asserted, in his second paragraph, that "no one who has examined the specimens exhibited by Mr. Prestwich and myself" doubted that they were human

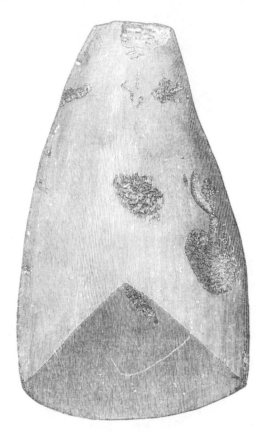

Figure 8. A typical polished-stone tool, of the type familiar to British archaeologists before 1858–59 (from Dawkins, *Early Man in Britain,* 1880).

artifacts. He declared himself confident that "on further examination Mr. Wright will see reason to change his opinion."[3] To facilitate such a change, Evans promised to leave several flint implements in the Society of Antiquaries' library for interested parties to inspect. The phenomena, he clearly implied, could speak more eloquently for themselves than he could speak for them.

Evans's appeal to the phenomena did nothing to change Wright's opinion. Wright found the specimens on display at the

3. John Evans, "Flint Implements in the Drift," *Athenaeum*, June 25, 1859, 841.

Society of Antiquaries no more convincing than those that Prestwich had shown to the Royal Society. He wrote a second letter to the *Athenaeum* on June 28, restating and amplifying his position. Even among what were presumably Evans's best specimens, Wright could find none that showed clear and distinct traces of human design.[4] This time it was Andrew Ramsay who answered Wright, in a letter dated July 13 and published in the July 16 *Athenaeum*.

Perhaps because Evans's appeal to the phenomena had failed to sway Wright, Ramsay appealed to authority instead. He began by noting that, in his twenty years as a geologist, he had "daily handled stones, whether fashioned by Nature or Art." Ramsay clearly implied that he and "others of his craft" were fully qualified to judge whether a stone had been shaped by nature or by human ingenuity. He went on to make clear his own position— that the flint implements from Abbeville were "as clearly works of Art as any Sheffield whittle"—and to suggest, accurately, that many of his colleagues shared that position. Ramsay gave reasons for his views but clearly believed that it was his expertise, and that of other geologists, that gave weight to the argument. This attitude surfaced in the last paragraph of the letter, where Ramsay discusses the position of the flint implements in the Drift. "I accept this part of the evidence from Mr. Prestwich alone, as I would accept the evidence of the existence of the planet Neptune from Prof. Adams. Mr. Adams's peers know his value, and all British, and most continental geologists are aware, that Mr. Prestwich is not only a man of long-tried experience, but is skilful and cautious in all his determinations."[5] Wright did not respond to Ramsay's letter. In the public eye, at least, the controversy ended with the geologists having the last, decisive word.

The third attack on the new evidence for human antiquity, though not backed by an impressive scientific reputation, took place in the most public of scientific forums. A day after Lyell's speech to the British Association for the Advancement of Science, Rev. John Anderson addressed the same geological section, bent on refuting Lyell's conclusions. Anderson, an amateur geologist

4. T. Wright, "Works of Art in the Drift," *Athenaeum*, July 9, 1859, 809.
5. A. C. Ramsay, "Works of Art in the Drift," *Athenaeum*, July 16, 1859, 83.

committed to reconciling science and scripture,[6] was well acquainted with the principal evidence for human antiquity. His
paper systematically examined the evidence, taking up Brixham
Cave, the Grotta di Maccagone, Abbeville, St. Acheul, and other
sites in turn. Anderson agreed that the chipped flints were crude
tools made by a race of savage humans and admitted that they
had been found among the bones of extinct animals. For each
site, however, he offered a geological interpretation that did not
challenge "the usually accepted date of man's very recent introduction upon the Earth."

Anderson argued that the association of human artifacts and
extinct animals was not proof that the two had once coexisted.
The stone tools and the bones had initially been deposited, he
claimed, in separate beds as relics of different, widely separated
periods. The collapse of cave roofs, the subsidence of the land
surface, and the undermining of sea cliffs eventually brought
them together in a single deposit. Such recent, accidental mixtures would be indistinguishable from deposits of great antiquity,
Anderson concluded, if they were cemented together by "petrify-
ing springs" or "silicious mud." Describing the audience's reaction
to Anderson's speech, Pengelly told his wife that "there was a
considerable amount of orthodoxy in the room, and he got a very
undue share of applause."[7]

The geologists in the British Association audience dealt with
Anderson's criticism as they had dealt with Wright's—politely,
firmly, and in detail. Pengelly described the scene this way: "Lyell
handled him as a gentleman and a philosopher alone can do it.
Next Phillips, having rubbed his hands in oil, smoothed him
down, but in such a way as to scarify him; then Ramsay seized
him by the button-hole and informed him of a fact or two connected with caverns, and finally handed him over to me, upon
which I seized him by the collar, dragged him into Brixham Cave,
and showed him its facts and their whereabouts."[8] All four of the
geologists who spoke were members of the human antiquity core

6. J. Anderson, *The Course of Creation* (1853).
7. W. Pengelly to Lydia Pengelly, September 17, 1859, in H. Pengelly, *A Memoir of William Pengelly* (1897), 90.
8. Ibid.

set. Their rebuttal to Anderson's arguments, like the answers to Wright in the *Athenaeum*, was less an attempt to change the mind of an individual critic than a public defense of the data on which their theories rested. Their belief that the defense had been successful is implicit in Pengelly's glowing report of the proceedings.

Wright had expressed doubts about the chipped flints' status as human artifacts. Anderson had challenged the genuineness of the tool-bone associations. In a letter written to the *Athenaeum* on November 8, 1859, Cambridge botany professor John Henslow questioned whether, at one well-publicized human antiquity site, such associations had existed at all. The site in question was Hoxne, to which both Prestwich and Evans had called attention in their papers. Henslow visited it in early November and interviewed two men who worked in the brick-earth quarries where stone tools had often been found. The older quarryman, Henslow reported in his letter, stated firmly that the tools occurred only a foot or two below the surface, well above the layer of earth that contained the bones of extinct animals. The younger man confirmed this testimony. Asked if the flints and bones had come from the same bed, he replied: "They must be very simple folk to think so. There have been many here to inquire, but they won't attend to what I have told them; they will have it otherwise."[9]

The sentences above—Henslow's only direct quotation from either witness—suggests that more than the status of the Hoxne evidence was at stake. They imply that Prestwich and Evans, among others, had come to Hoxne and seen in its strata only what they wanted to see. The geologists, according to Henslow's thinly veiled accusation, had been so convinced of the truth of their theories that they had ignored the testimony of a reliable eyewitnesses, the quarrymen. In a scientific community that held empiricism sacred, this was a serious charge.

Prestwich answered the charge in a letter dated December 1 and published in the December 3 *Athenaeum*, two weeks after Henslow's letter had appeared. The specifics of his answer were less important than the care that he took to answer each of Henslow's objections in exacting detail. Prestwich cited the strati-

9. J. S. Henslow, "Celts in the Drift," *Athenaeum*, November 19, 1859, 668.

graphic differences between various sections of the quarry and
the potentially confusing names that the quarrymen applied to
various beds. He also argued that in several cases the quarrymen's
information had been imprecise or simply wrong. In one in-
stance, the younger men had told Henslow that two stone tools
had been found one or two feet below the surface. Prestwich,
when shown the place where they had been found, estimated
that they had been "at least" eight to ten feet below the surface.[10]
Throughout his response to Henslow, Prestwich emphasized his
own observations at Hoxne, which he regarded as a necessary
supplement to and check on the observations of the quarrymen.

Prestwich wrote a second letter, within a week of the first,
that expanded on the need for firsthand observation. Published
in the December 10 *Athenaeum*, the letter offered a defense, detailed summary of the stratigraphic information Prestwich had
collected on his second visit to Hoxne, in early fall. The letter was
an aggressively empiricist document—paragraphs of geological
data relieved by very little interpretive prose. Its last sentence
clearly stated Prestwich's faith in empiricism, cautioning that "a
point like this, involving the question of the association in the
same deposit of the works of man and the remains of the extinct
Elephant and other animals, can only be satisfactorily settled by
positive facts and accurate observations."[11]

Two weeks later, in a letter printed in the December 24 *Athenaeum*, Henslow conceded that Prestwich's "remarks on the statement of the old man at Hoxne pit are very likely just." The stone
tools, Henslow went on, had probably come from the fossiliferous
beds that Prestwich (and, in 1797, John Frere) believed they had.
Henslow counseled caution, noting that some stratigraphic questions about Hoxne were still unresolved, and that the exact age
of the stone tools was uncertain. Nevertheless, he agreed that "if
the observations at Abbeville and Amiens have been satisfactorily
confirmed, there can remain no doubt with any geologist that the
Hoxne pit will very probably prove an analogous case."[12] Hen-

10. J. Prestwich, "Flint Implements in the Drift," *Athenaeum*, December 3,
1859, 740–41.
11. J. Prestwich, "Flint Implements in the Drift," *Athenaeum*, December 10,
1859, 775–76.
12. J. S. Henslow, "Works of Art in the Drift," *Athenaeum*, December 24,
1859, 853.

slow thus conceded the most important point in the debate to Prestwich. The stone tools *had* been found alongside the Post-Pliocene mammal bones; the geologists had not been led astray by their theories.

Gathering Corroborative Data, July 1859–December 1862

The case for human antiquity that Prestwich and his colleagues presented in mid-1859 rested on evidence from Abbeville and St. Acheul, corroborated by discoveries at Brixham Cave, the Grotta di Maccagone, and Hoxne. Members of the core set had carefully investigated each site and agreed that each had yielded solid evidence for human antiquity. The members of the core set also agreed, however, that the case would be significantly stronger if they corroborated it with additional data from new sites (fig. 9). The geologists' belief that more data would make a sounder theoretical case reflected their commitment to empiricism, but it had a practical dimension as well. The core set's case for human antiquity, if based on a wide range of sites, would not be seriously damaged if subsequent discoveries discredited one of the original sites.

Intensive investigations of sites that might yield corroborative data began in the summer of 1859 and continued until Lyell's *Antiquity of Man* appeared in January 1863. The investigations were carried out primarily by geologists but were fundamentally different from the geological investigations carried on at Brixham Cave and in the Somme Valley. Rather than intensively studying a few sites, the new investigations surveyed promising human antiquity sites throughout France and Britain. The goal of the new investigations was different as well: the collection of stratigraphic and paleontological data to support a theory, rather than the formation of the theory itself. Prestwich and his colleagues worked, therefore, as individuals rather than as members of a core set. They integrated their search for new human antiquity sites into their ongoing, individual research programs.[13]

Prestwich, in particular, searched enthusiastically for corrob-

---

13. See, e.g., Prestwich's letters to Lyell on December 28, 1859, January 14, 1860, and September 16, 1861, Lyell Papers, EUL. W. Pengelly to Lyell, February 14, 1861, and September 9, 1862, are also illustrative, as are Falconer's letters to Lyell, Lyell Papers, EUL, and Prestwich, Falconer Papers.

orative data. In late September and early October 1859, he and Falconer explored a number of caves in the Gower Peninsula of Wales. The results of the trip–discoveries of elephant, rhinoceros, hyena, and bear remains as well as flint tools—formed the subject matter of a paper Falconer read to the Geological Society the following spring.[14] In December, 1859, Prestwich wrote to Lyell to discuss a rhinoceros fossil that had been found in the tool-bearing beds of Menchecourt in the 1840s.[15] He made several visits to France during 1860, including a field trip in April on which he was accompanied by anatomist George Busk and archaeologist John Lubbock. Prestwich and Lubbock returned to France in April 1861, accompanied by John Evans. Soon after their return, the three departed for Bedford, England, in order to examine yet another new human antiquity site. They visited France again in 1862 and 1863, trips that provided Lubbock with much of the data for his book *Pre-historic Times* (1865).[16]

The restructuring of the human antiquity investigation during the summer of 1859 made it possible for amateur geologists to play a much larger role than they had before. The focus of the investigation had shifted from theoretical discussion, the province of expert geologists, to data collection, a process in which local knowledge could be vital. In addition, the publicity attracted by the experts' mid-1859 papers excited amateur interest in the human antiquity question. Amateurs throughout Britain, along with some in France, searched for new examples of flint implements in the Drift and announced their discoveries in print or in letters to expert geologists. On February 23, 1861, for example, Prestwich and Evans traveled to the North Sea coast, where several flint implements had been found between Herne Bay and Reculvers. Prestwich had learned of the discovery from a local geologist named Thomas Leech.[17] The discovery of "two fine hatchets of the true Amiens and Hoxne type" by a local newspaper editor named Wyatt brought not only Prestwich, Evans, and

14. G. Prestwich, *Life and Letters of Sir Joseph Prestwich* (1899), 139–40.

15. Prestwich to Lyell, December 18, 1859, Lyell Papers, EUL.

16. G. Prestwich, *Life and Letters*, 147, 154; Evans to J. Prestwich, April 18, 1861, in G. Prestwich, *Life and Letters*, 163–65.

17. J. Prestwich to John Evans, 5 January 1861, in G. Prestwich, *Life and Letters*, 162.

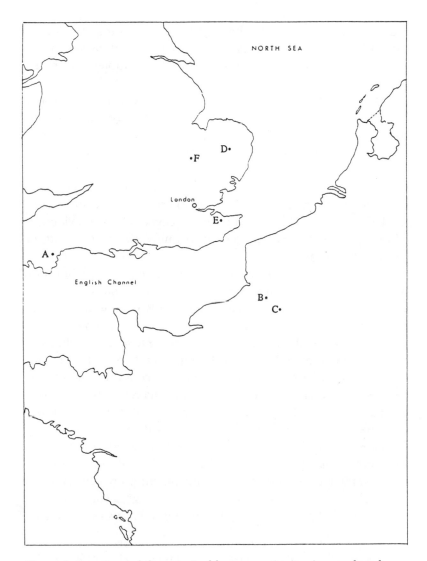

Figure 9. Locations of the principal human antiquity sites explored by British geologists between 1858 and 1863. *A*. Torquay (Brixham Cave, Kent's Cavern); *B*. Abbeville (Menchecourt, Moulin-Quignon); *C*. Amiens (St. Acheul, St. Roch); *D*. Hoxne; *E*. Herne Bay and Reculvers; *F*. Bedford.

Lubbock but also Lyell to the important new site at Bedford.[18] William Brockie, another amateur, wrote to the *Athenaeum* in late 1859 to announce the discovery of several flint implements in his home town, South Shields.[19]

Not all searches for corroborative evidence took place in the field. As proof of human antiquity accumulated, geologists began to reassess the work of Schmerling, MacEnery, Tournal, and others. John Evans noted, in 1861, that doubts about the validity of cave evidence "are now in great degree dispelled by similar discoveries having been made under circumstances which preclude the interference of those causes of error which come within the bounds of possibility in the case of caverns." Cave evidence, Evans concluded, now had "a right to a new trial at the hands of scientific inquirers."[20]

Geologists led the search for corroborative evidence, but they were not the only scientists involved. Stirred, perhaps, by the investigation's new empirical character, the British archaeological community joined the search. The historical archaeologists refused to theorize about prehistoric Britons but saw the collection of stone tools as a form of fact gathering. Members of the Archaeological Institute examined a "considerable collection of antiquities of stone, weapons, implements, and objects of unknown use" at their April 13, 1860, meeting. The collection included more than twenty stone tools, exhibited by a half-dozen archaeologists, from a variety of English and Irish sites.[21] Members of the British Archaeological Association and the Ethnological Society exhibited a similar collection at their joint meeting on February 19, 1861. The meeting's organizer, Thomas Wright, concluded his opening remarks by "recanting his former opinions as to the nat-

18. Lyell to Sir Charles Bunbury, April 26, 1861, in K. Lyell, *Life and Letters of Sir Charles Lyell* (1881), 2:344. Also see John Evans to J. Prestwich, April 18, 1861, and J. Prestwich to John Evans, "Thursday," in G. Prestwich, *Life and Letters,* 163–64. Evans described the Bedford and Herne Bay-Reculvers implements, along with a number of others, in "Account of some Further Discoveries of Flint Implements," *Arch.* 39 (1862): 57–84.

19. W. Brockie, "More Flint Implements," *Athenaeum,* December 17, 1859, 815.

20. John Evans, "Further Discoveries," 78.

21. "Proceedings at Meetings of the Archaeological Institute," *AJ* 17 (1860): 169–72.

ural character of these fossil implements, and admitting his subsequent conviction that they were really of human manufacture."[22]

The presence of both leading geologists and prominent archaeologists at these meetings demonstrates that the flow of information between experts and amateurs ran both ways after the summer of 1859. Godwin-Austen displayed and discussed a group of stone tools from the Somme Valley at the 1860 meeting. Evans gave a similar presentation at the joint 1861 meeting, and Pengelly followed it with a summary of the data from Brixham Cave. These presentations served a dual purpose. First, they provided amateur enthusiasts, who might never have inspected a chipped-flint implement firsthand, with a clear idea of what to look for. Second, they provided those who had not heard the 1859 papers with an explanation of the tools' geological context and a summary of the geologists' consensus on human antiquity. These same motivations led Prestwich to send two stone tools collected in Abbeville to Robert Chambers in November 1859. Chambers, a leading publisher with strong scientific interests, enthusiastically supported human antiquity and promised to show the tools to a "large and intelligent audience" at an upcoming meeting of the Edinburgh Philosophical Institute.[23]

The process of collecting more data on human antiquity involved large segments of the British scientific community. It made the details of the human antiquity question intimately familiar to ethnologists, archaeologists, and amateur geologists whose interest had been excited by the 1859 papers. In doing so, it created ties between the geological, archaeological, and ethnological communities, beginning a tradition of joint research on human antiquity that lasted until the end of the century. It also had shorter-term effects. The extensive data-collection and publicity campaign that began in summer 1859 laid the groundwork for the geological community's last wholly independent contributions to human antiquity research.

22. "[Proceedings of] Archaeological and Ethnological Societies," *Geologist* 4 (1861): 153.
23. G. Prestwich, *Life and Letters,* 142.

The Problem of Human Fossils

The three-year search for evidence to corroborate the 1858–59 discoveries uncovered hundreds of new flint implements and added dozens of new sites to the list of those with well-documented tool-bone associations. It failed, however, to close a large and frustrating gap in the new case for human antiquity. Three years of scouring the Post-Pliocene strata of Western Europe had disclosed an abundance of stone tools but no sign of the toolmakers. The principal human antiquity sites were rich in animal remains, but destitute of even a single human bone. The experienced geologists who led the search found this both expected and troubling.

Fossilized human remains had always been scarce. They had been discovered in Europe only four times since the beginning of the nineteenth century. Philippe-Charles Schmerling's excavations of Engis, Engihoul, Chokier, and other Belgian caverns in the 1830s had yielded two human skulls and a variety of other bones. A nearly complete human skeleton had been unearthed in 1844 from the Post-Pliocene volcanic deposits near Denise, France. The fossil remains of seventeen individuals had been recovered from a cave near the French village of Aurignac in 1852, only to be reburied in an unmarked churchyard grave a few days later. Finally, in 1857, the celebrated remains of "Neanderthal Man" had been discovered in a cave high above the banks of the Neander River, near Düsseldorf, Germany. Though they were quickly brought before German scientific authorities, the Neanderthal bones were not well known in Britain until 1861, when the *Natural History Review* published a detailed description of them.[24]

Three of these four finds were accidental. Only Schmerling's was the product of an organized search. Since 1857, however, the discovery of scores of stone tools in Post-Pliocene caves and gravel beds suggested that humans had been numerous in Post-Pliocene Europe. An intensive search, focused on sites known to contain (or likely to contain) stone tools should therefore have uncovered a significant number of human bones. The fact that it

24. J. Reader, *Missing Links,* 2d ed. (1988), 7–8.

had not implied—to critics as well as to some supporters—that humans had been far less common in Post-Pliocene Europe than geologists suggested. Combined with lingering suspicions that the "stone tools" might be accidents of nature, the dearth of human fossils allowed critics to suggest that humans had never been there at all.[25]

The geologists' inability to locate human fossils in the Drift was frustrating for another reason, as well. It deprived them of what should have been their best, most conclusive evidence for human antiquity. A human bone, removed from undisturbed deposits that also contained stone tools and the remains of extinct animals, would have provided the scientific equivalent of a lawyer's "smoking gun": irrefutable evidence that men had lived among the mammoths. Most scientists were willing, by the early 1860s, to accept the coexistence of humans and Post-Pliocene animals on the evidence of stone tools alone, but a human fossil would have done much to convince skeptical members of the general public.

None of the four sets of human fossils discovered earlier in the century provided adequate proof of human antiquity. The volcanic beds of Denise contained no bones of extinct animals, and no stone tools—their Post-Pliocene age could be demonstrated only by a complex chain of geological reasoning. The other three sets of fossils came from caves, and so were tainted by caves' reputation for complex stratigraphy and contents disturbed by natural processes.[26] Each of the three cave sites also had unique flaws of its own. Quarrying had destroyed several of Schmerling's most productive caves including Engis, where he had found two human skulls.[27] The Aurignac skeletons were lost in an unmarked grave, and the 1852 excavations, carried out by amateurs, had badly disturbed the cave strata. Furthermore,

25. See, e.g., "Review of *Geological Gossip*," *British Quarterly Review* 32 (1860): 258–59; and W. H. Smith, "Wilson's *Prehistoric Man*," *Blackwood's* 93 (1863): 526–27.

26. The most notorious example of the potential of cave deposits to mislead unwary investigators was William Buckland's investigation (and dismissal) of an alleged human fossil, the "Red Lady of Paviland"; See F. J. North, "Paviland Cave," *Annals of Science* 5 (1942): 91–128.

27. C. Lyell, *Geological Evidences of the Antiquity of Man* (1863), 69.

as Lyell pointed out in an 1861 letter to French paleontologist Edouard Lartet, the artifacts found in the cave apparently belonged to a culture more advanced than that responsible for the stone tools of the Drift.[28]

Geologists could not date the Neanderthal skeleton with any certainty. The loam in which it lay contained no animal bones, and could not be connected with any strata of known age. The skull's primitive appearance was, in itself, no proof of great antiquity. Claiming that the skull's heavy brow ridges, low crown, and protruding face marked Neanderthal Man as an ancient ancestor of modern humans meant accepting the controversial theory of humankind's evolution from the lower animals. Many believed that the same features could be more economically explained as the results of old injuries or degenerative diseases.[29] The Neanderthal skeleton was unquestionably an interesting discovery, but in the absence of similar skeletons from better-dated sites, it added nothing to the debate over the age of the human race.

Boucher de Perthes had searched the Somme River gravels for thirty years without finding a human fossil. He had long predicted, however, that the gravels would eventually yield one. His declaration, in the early spring of 1863, that he had recovered a fossilized human jaw and teeth from Moulin-Quignon seemed to not only to fulfill his prophecy but to close, in a single stroke, one of the most troubling gaps in the new case for human antiquity. The circumstances of the find were everything that might have been hoped for. Moulin-Quignon was not a cave site, and the presence of both stone tools and Post-Pliocene animal bones in its gravel beds had been well documented since Evans and Prestwich's first visit to the area in 1859. The workmen who first found the jaw had disturbed neither it nor the now-familiar chipped stone tools found with it.

It soon became clear, however, that the Moulin-Quignon jaw was not all that it appeared to be. Elation over the discovery of such a compelling piece of evidence quickly turned to suspicions and in time to accusations of fraud. The jaw's value as evidence

28. Lyell to Lartet, November 22, 1861, in L. G. Wilson, *Charles Lyell's Scientific Journals on the Species Question* (1970), 527–30.

29. Reader, *Missing Links,* chap. 2.

plummeted, but in the long run, the role that British geologists played in the controversy not only enhanced their reputations but also strengthened their case for human antiquity.

The Moulin-Quignon controversy began in a rush of excitement on March 23, 1863.[30] Workmen digging in the gravel beds at Moulin-Quignon, near Abbeville, brought Boucher de Perthes two flint tools and a single human molar. Five days later, the workmen called de Perthes to the same pit and showed him half of a human lower jaw, still imbedded in the matrix and with one tooth attached. De Perthes extracted the partial jaw from the gravel. A second amateur geologist, who had come to the pit with him, found a flint hatchet and a second tooth in a nearby section of the same gravel bed. News of the discoveries was greeted with delight not only in London but also in Paris, where most geologists and archaeologists had, by late 1859 or 1860, accepted that humans had lived in Post-Pliocene Europe.[31]

Prestwich, Evans, and Falconer—the three most active figures in the human antiquity investigation—went to Abbeville in mid-April to see the new finds firsthand. Their initial reaction was mixed. Prestwich and Evans examined the tools on April 13, and immediately suspected them of being modern forgeries. Falconer, who arrived the next day, concurred that the tools were forgeries, but provisionally concluded that the jaw itself was both genuinely human and extremely old. It possessed, he told British zoologist William B. Carpenter and French zoologist Armand de Quatrefages, a suite of characteristics unlike those of any modern European race. This "primitive" appearance effectively ruled out any chance that the jaw belonged to a Celt or a Roman living long after the Post-Pliocene gravel beds were deposited. De Perthes had apparently discovered the remains of a genuine, prehistoric flint-tool-maker.[32]

---

30. H. Falconer et al. ("Proceedings of the Late Conference," *Natural History Review* 3 [1863]: 432–63) recount the facts of the Moulin-Quignon affair at length. The best summary is D. K. Grayson, *The Establishment of Human Antiquity* (1983), 213–17.

31. Grayson, *Human Antiquity*, 190–95. One notable exception to the French enthusiasm for human antiquity was Élie de Beaumont, a geologist who, like Lyell, enjoyed an international reputation.

32. Grayson, *Human Antiquity*, 213–14.

English enthusiasm over the Moulin-Quignon jaw lasted no more than a week. Falconer returned to London on April 18, carrying one of the loose teeth and a cast of the jaw with him. He examined them carefully during the next four days—calling Prestwich, Evans, and anatomist George Busk in as consultants—and concluded that neither the jaw nor the tooth was ancient. The jaw's morphology was, Falconer discovered, unusual but neither unique nor primitive. In fact, there were similar jaws in a large collection of modern skulls that had been found in a London churchyard some years before. Now convinced the jaw and teeth, like the tools, had been fraudulently introduced into the Moulin-Quignon gravels, Falconer wrote a long letter to the London *Times* explaining his position. Events outran him, however. By the time Falconer's letter appeared on April 25, Carpenter and Quatrefages had submitted papers to the Royal Society and the Académie des Sciences, respectively, announcing the jaw's authenticity and citing Falconer's original opinion as evidence.[33]

Battle lines were drawn by early May, the English claiming that both the jaw and the tools were modern additions to the gravel, the French claiming that tools, bones, and gravel were all of Post-Pliocene age. A five-day Anglo-French conference, designed to settle the issue, convened first in Paris and then in Abbeville but changed very few minds. The French continued to believe that the jaw was authentic and, after discovering genuine stone tools at Moulin-Quignon, Prestwich temporarily adopted their position. The conferees' official report stated that the jaw had not been fraudulently introduced, that most of the stone tools found with it were genuine, and that both were contemporary with the deposition of the gravel. Busk and Falconer remained unconvinced, but they were outvoted by Prestwich and the French.[34]

Evans, who had missed the conference, remained both unconvinced and free of any obligation to publicly support the con-

---

33. P. Boylan, "The Controversy of the Moulin-Quignon Jaw," in *Images of the Earth*, ed. L. J. Jordanova and R. Porter (1979). Boylan's detailed analysis is based in part on Falconer's letters.

34. The *"procès-verbaux"* of the conference are printed in French with annotations in English, in Falconer et al., "Proceedings of the Late Conference," 431–62.

ference's official findings. He asked Henry Keeping, the fossil collector who had worked under Pengelly at Brixham Cave, to go to Moulin-Quignon and examine the site for himself. Keeping arrived in Abbeville on June 3 and, with de Perthes's blessing, took part in the ongoing excavation of the area where the jaw had been found. Between June 3 and June 7, Keeping found seven stone tools at Moulin-Quignon, all of which had been recently inserted in the gravel. "I have," he reported to Evans, "every reason to believe that all the specimens I have brought from Moulin-Quignon were placed there on purpose for me to find."[35] This testimony confirmed the British scientists' suspicions, and brought Prestwich back into their camp, destroying the last traces of British support for the Moulin-Quignon finds.

The British geological community's reaction to the Moulin-Quignon episode was particularly significant. Falconer, Evans, Prestwich, and Alfred Tylor were uniformly skeptical of the jaw's geological position and of the flints originally found alongside the jaw. As each of the four suspected fraud, he quickly made his opinion public and worked vehemently to discredit the jaw as evidence of human antiquity.[36] The British geologists' denunciations of the fraud all placed blame on Boucher de Perthes's hired workmen. In a typical response, Tylor argued that "years of practice in fabricating sham antiquities, with the additional stimulant of the reward offered by M. Boucher de Perthes, have at length enabled the quarrymen to put bones and implements into the gravel so skilfully as to deceive even the Patriarch of Primeval Archaeology himself."[37] Tylor left no doubt that the extent of the quarrymen's cunning, not Boucher de Perthes's shortcomings as a scientist, accounted for the fraud's success.

The British geologists' responses to the Moulin-Quignon evidence reflected their belief that the credibility of their case for

35. H. Keeping, "Flint Implements Found at Moulin-Quignon," quoted in John Evans, "Human Remains at Abbeville," *Athenaeum,* July 4, 1863, 19–20.

36. H. Falconer, "Falconer on the Reputed Fossil Man of Abbeville," *Anthropological Review* 1 (1863): 177–79. A. Tylor, "On the Discovery of Supposed Human Remains," *Anthropological Review* 1 (1863): 166–68; John Evans, "Abbeville Human Jaw," *Athenaeum,* June 6, 1863, 747–48; Evans, "The Human Remains at Abbeville," 19–20.

37. Tylor, "Discovery of Supposed Human Remains," 168.

human antiquity was at stake. By moving quickly to denounce the fraud, they demonstrated that they could readily tell genuine artifacts from forgeries and real fossil bones from "planted" ones. By denouncing a human fossil—a type of evidence that they had waited years to find—they showed that their observations were not prejudiced by theory. The British geologists' attack on the Moulin-Quignon jaw thus prevented the forgery from damaging their case while simultaneously enhancing their reputations as observers. Falconer's April 25 letter to the *Times*, for example, ended with the statement, "The breakdown in this spurious case in no wise affects the value of the real evidence, now well established." [38] Evans's June 6 letter to the *Athenaeum* noted that he and his colleagues had denounced the forgery even though it would have supported their case. Pengelly, addressing the Torquay Natural History Society in December, stressed both points. He concluded that the advocates of human antiquity "have not been tripped up by their opponents. The fraud, if it be one, was detected and proclaimed to the world as such by the very men who were and are amongst the foremost to teach that the advent of man took place at a period vastly more remote than has commonly been believed." [39]

How much damage could the Moulin-Quignon episode have done to the case for human antiquity? A century and a quarter after the fact it is impossible to be certain, but circumstantial evidence suggests that the damage might have been extensive. Forged artifacts were part of archaeology, familiar both to scientists and to the public, long before the age of the human race became an issue. Stone tools were, in fact, among the easiest artifacts to duplicate; Evans estimated that a skilled workman could make one in less than fifteen minutes. [40] When the discoveries at Brixham Cave were announced, charges that the investigators had been taken in by forgeries quickly followed. [41] Falconer, at least, believed that exposure of the Moulin-Quignon fraud "by

---

38. Falconer, "Fossil Man of Abbeville," 179.
39. W. Pengelly, *Introductory Address Delivered before the Torquay Natural History Society* (1864).
40. Evans to Lyell, June 18, 1863, Lyell Papers, EUL.
41. Falconer to Pengelly, October 17, 1858, Brixham Notebook 10.

the enemy," would have given similar charges against the geologists a factual basis and so been a serious setback for the geologists' cause.[42]

Lyell's Antiquity of Man

Even as the Moulin-Quignon affair took shape, Charles Lyell published *Geological Evidences of the Antiquity of Man*, a substantial book that presented the geologists' case for human antiquity as a fait accompli. Though written by a single author, *Antiquity of Man* was substantively and symbolically the work of all the human antiquity investigators. It drew extensively on Prestwich's and Evans's work in the Somme Valley, Pengelly's excavation of Brixham Cave, Falconer's investigations of caves throughout Europe, and a variety of recent archaeological studies. It combined data from recently investigated sites with data collected decades before by Schmerling, Tournal, and Boucher de Perthes and used both to argue for the presence of humans in Post-Pliocene Europe. Lyell's book was not overtly theoretical, but his belief in the antiquity of the human race was unmistakable. His knowledge of post-Tertiary geology and his reputation as theorist meant that, in expressing his belief in men among the mammoths, he tacitly spoke for all the geologists working on the human antiquity question. *The Antiquity of Man* was, in this respect, Lyell's 1859 British Association for the Advancement of Science address on a grand scale.

The book appeared in three editions during 1863: the first in February, the second in April, and the third in November. The differences between editions were, as Donald Grayson's careful analysis has shown, matters of a few words or at most a few sentences. The changes were principally responses to charges, by Falconer and Prestwich, that Lyell had slighted them.[43] The changes

42. Falconer to Grace McCall, 28 April 1864, Falconer Papers, 117.

43. W. F. Bynum, "Charles Lyell's *Antiquity of Man* and Its Critics," *Journal of the History of Biology* 17 (1984): 153–87. "I charge him," Falconer wrote privately to his niece, "with appropriation of my labours without acknowledgement; and of systematic disparagement of my discoveries," Falconer to Grace McCall, February 12, 1863, Falconer Papers, 93.

Lyell also had numerous defenders within the scientific community. Pengelly took issue with Falconer's charges in two strongly worded letters to Lyell

that incorporated newly discovered facts or newly formed theories did not affect the structure of Lyell's argument or his claims for human antiquity.[44] I will, therefore, treat the three editions as a single work and quote only from the first.

Lyell's review of the evidence for human antiquity began, as his earlier treatments of geological time had begun, in the present. The second and third chapters of his book concerned human remains and artifacts from the Recent epoch: those found in Danish peat bogs and shell mounds, in Swiss lake-dwelling sites, and in the valleys of the Nile, Mississippi, and Ohio rivers. The next eight chapters (4 to 11) dealt with evidence from the earlier, Post-Pliocene epoch: Schmerling's discoveries in Belgium, the Neanderthal skeleton, the tools and bones from the Somme Valley sites, and subsequent discoveries in alluvial deposits and caves. In chapter 12, Lyell stepped back in time again, to the glacial period that coincided with the late Pliocene and early Post-Pliocene. The end of the glacial period, he argued in chapters twelve through eighteen, marked the beginning of a human presence in Europe. No human remains—bones or artifacts—had yet been found in or below the Boulder Clay stratum deposited by the glaciers and drifting ice of the early Post-Pliocene.

The nineteenth chapter of *Antiquity of Man* summarized the evidence in the first eighteen. It did not, however, bring the book to a close. Lyell devoted his last five chapters to a discussion of theories of transmutation in general and Darwin's theory in particular, paying close attention to the implications for human biological and cultural development. These chapters, so cautious that they irritated Darwin,[45] annoyed many contemporary reviewers

---

(Pengelly to Lyell April 11 and 13, 1863, Lyell Papers, EUL), and geologist George Poulett Scrope praised Lyell's own published response to the charges (Lyell to Scrope, April 29, 1863, Lyell Papers, American Philosophical Society [APS], B L981.16. Alfred Russel Wallace believed that Falconer's letter was "quite uncalled for" and assured Lyell that "the general impression on every impartial reader of your book must have been, that you had endeavoured to do full justice to all parties whose observations you had occasion to mention" (Wallace to Lyell, April 22, 1863, APS Miscellaneous Papers, B D25.L).

44. D. K. Grayson, "The First Three Editions of Charles Lyell's *Geological Evidences of the Antiquity of Man*," *Archives of Natural History* 13 (1985): 105–21.

45. Bynum, "*Antiquity of Man* and Its Critics," 157.

simply by being part of the book. *Antiquity of Man* was widely reviewed, and the charge that Lyell had devoted too much space to the glacial period and to the origin of species was a common refrain among its reviewers. One anonymous reviewer concluded that Lyell would have done better to compile "a small work purely on the geological evidence of the antiquity of man." Another, writing in the *Westminster Review*, sought to separate Lyell's argument for human antiquity from the "great deal of irrelevant matter" found in the book.[46]

Nineteenth- and twentieth-century commentators alike have asserted that *Antiquity of Man* lacked thematic coherence, but this judgment seriously undervalues the book.[47] It is true that, despite its title, *Geological Evidences of the Antiquity of Man* was neither entirely geological nor entirely concerned with the human antiquity question. It is also true, however, that Lyell's book was unique among contemporary scientific works in the geographic and disciplinary breadth of its synthesis. Earlier writers on human antiquity—even those who, like de Perthes, had produced books—had focused on a single site or a group of neighboring sites. Lyell discussed sites ranging from the Nile Delta to the North Sea and from Central Europe to the Mississippi River, and he unified data from many areas under a single hypothesis. Lyell also reached beyond the intellectual frontiers of geology, drawing connections between the geological evidence for human antiquity and the data collected by archaeologists, anatomists, and linguists.

In the second and third chapters of *Antiquity of Man*, Lyell discussed the relics of ancient Egyptian civilization and of the European Bronze and Iron ages. By discussing pottery from the Nile Valley, iron swords from Switzerland, and bone tools from Danish shell heaps, he demonstrated that humans had lived in Europe not for the few thousand years allotted in Genesis but for tens of thousands of years. He also demonstrated that a clear connection existed between the artifacts studied by archaeologists

46. "Lyell on the Geological Evidence of the Antiquity of Man," *Anthropological Review* 1 (1863): 136; "The Antiquity of Man," *Westminster Review* 79 (1863): 274.

47. In addition to the Victorian reviewers already cited, see Bynum, "*Antiquity of Man* and Its Critics," 157, 183–85.

and the primitive tools that geologists had discovered in the Drift. Lyell's later chapters on the glacial period served a similar dual purpose. On one hand, they established the lower limit of humanity's time in Europe; on the other hand, they showed how studies of early humans merged with the traditional subject matter of geology. The last chapters of *Antiquity of Man,* which sought to link the age of the human race with the development of human language, human races, and the human species itself, attempted—albeit only partially successfully—to tie together geology and anthropology.

W. F. Bynum has contended that Lyell's discussions of anthropological issues were highly derivative and limited by Lyell's unfamiliarity with the data of anthropology. This is an accurate assessment. For example, Lyell's discussion of the T. H. Huxley–Richard Owen debate over the similarities of human and simian brains is little more than a statement of Huxley's position. *Antiquity of Man* was not, however, primarily an anthropological work. The majority of its chapters offer an original synthesis of purely geological data. Though the principal subject of those chapters was man, their purpose was geological: locating the earliest Europeans among the faunas and geological events of the late Tertiary and Quaternary periods. Several commentators, in fact, made a point of praising Lyell's synthetic treatment of Quaternary geology.[48]

Lyell began his geological career in the 1820s, when the science had rigorously excluded humans from its subject matter except as a marker of the end of geologic time. He knew, from firsthand studies of the data, that the discoveries made since 1858 rendered such an exclusion obsolete. Artificial separation of the study of geology from the study of early human history was no longer justifiable in 1863. Lyell's discussion of anthropological issues in the last chapters of the book, like his discussion of archaeological data in its first chapters, was an attempt to illuminate the links between geology and what were quickly becoming allied sciences. Lyell was not alone in this enterprise. His argument that archaeological and geological studies formed a continuum

48. See, e.g., Wallace to Lyell, April 22, 1863, APS Miscellaneous Papers, B D25.L.

echoed the opening words of John Evans's June 1859 paper on the stone tools from Abbeville and Amiens and foreshadowed the synthesis of geological and archaeological ideas that emerged in John Lubbock's 1865 book *Pre-historic Times*.

Lyell's *Antiquity of Man* was to become one of the founding documents of a new approach to archaeology. It influenced young geologists such as James Geikie and William Boyd Dawkins, much as *Principles of Geology* had influenced an earlier generation. Most readers, though, agreed with Darwin's opinion of the book: "a compilation . . . of the highest class."[49] Lyell's masterful presentation of the evidence for human antiquity further broadened discussion of the human antiquity question, raising new issues and drawing scores of educated Victorians into an already lively debate.

49. C. Darwin to J. D. Hooker, February 24, 1863, in F. Darwin, *Life and Letters of Charles Darwin* (1887), 3:8.

SIX

The Public Debate over Human Antiquity, 1859–75

THE NEW CASE for human antiquity was created between 1859 and 1863—a period of tumultuous intellectual change in Britain. Darwin's *Origin of Species,* published less than two months after Lyell delivered his landmark address on human antiquity to the British Association for the Advancement of Science, threatened to overturn long-held ideas about the history of life, God's role in that history, and, by extension, God's place in the natural world. Within five years, the work of T. H. Huxley and Alfred Russel Wallace expanded Darwin's theories to include humans as well as the lower animals. The idea that humankind might have descended, by natural processes, from an apish ancestor lay, as Alvar Ellegard has shown, at the heart of Victorians' distress over evolutionary theory. Liberal members of the clergy mounted an equally strong challenge to traditional beliefs. Bishop J. W. Colenso and the authors of *Essays and Reviews* (1860) argued that the Bible should be seen not as the directly revealed word of God but as the work of fallible human authors.

The new case for human antiquity raised many of the same issues, and touched many of the same nerves, as works on evolution and the new biblical criticism. Scientists, clergymen, and lay people alike recognized that questions about humankind's age were closely related to questions about its origins. Evolution demanded time, and if the first humans appeared only recently they must have done so by nonevolutionary means. The presence of men among the mammoths—the foundation of the new case for human antiquity—challenged ideas about "man's place in nature" already under attack by evolutionists such as Huxley. The idea that humans had lived on Earth far longer than 6,000 years contradicted traditional readings of Genesis, further corroding belief in the accuracy and inspired nature of the Bible.

Ties to high-stakes scientific and religious controversies had a paradoxical impact on the new case for human antiquity. The ties heightened public awareness of the human antiquity question, and spurred a wide-ranging public debate over the geologists' new solution and its implications. At the same time, the controversies surrounding evolution and biblical criticism often overshadowed the debate over human antiquity.

The new case for human antiquity threatened orthodox views of the past, of man's place in nature, and of the meaning of Genesis. Its threat paled, however, beside those raised by the liberal clergy and the evolutionists. Claims that some chapters of Genesis might need reinterpretation were troubling, but claims that the Bible was not inspired struck at the foundations of Christianity. Theories that humans had lived in Post-Pliocene Europe might entail a new view of the Divine Plan underlying the history of life, but *Origin of Species* denied that there was such a plan. The new case for human antiquity was the last threatening of several challenges to established ideas about the past and the easiest to reconcile with those ideas. The coexistence of humans and extinct animals had been well established by 1863 on the basis of fossil evidence. The age of the oldest humans, however, was still widely disputed. Geologists offered a broad spectrum of opinions, and other commentators freely adopted whichever one best fitted their existing ideas about nature.

The timing and flexibility of the new case for human antiquity minimized the distress it caused all but the most conservative Victorians. The new case changed ideas about human origins, "man's place in nature," and the meaning of Genesis but did so in subtle and largely nonrevolutionary ways. The commentaries on human antiquity published between 1859 and 1875 reveal the wide variety of ways in which educated Victorians came to terms with geologists' new picture of the recent past.

Discussions of important scientific ideas did not, in mid-Victorian Britain, take place solely or even primarily within scientific societies. New theories were constructed and debated in private, and formally presented at society meetings, but the implications of those theories were often debated in the wider and more public forum provided by the periodical press. Darwin's theory of evolution by natural selection was, not surprisingly, the most widely (and heatedly) discussed aspect of mid-Victorian science, but other major theories were not neglected.[1] During the 1860s and 1870s, Britain's leading general-interest reviews also discussed the glacial theory, the age of the Earth, and a variety of topics from the emerging science of prehistoric archaeology.

The existence of what R. M. Young has called a "common context" made these wide-ranging discussions possible. Career scientists, clergymen, and members of the educated public shared a familiarity with leading scientific theories and methods, an interest in the details of natural history, and a commitment to particular views of God's relationship to nature. This common intellectual context did not create a completely level playing field: Scientists enjoyed a privileged position, particularly when discussing the validity of a given theory. The existence of a common context did mean, however, that scientists and nonscientists alike could discuss a new discovery's implications on something like equal terms. It also meant that most educated Victorians could readily follow, and even participate in, the scientific discussions that appeared in the periodical press.[2]

The age of the human race, like the mode of origin of species, became the center of a long, wide-ranging discussion in the British periodical press. The first brief commentaries on the new case for human antiquity appeared in late 1859, and the first full article appeared a year later. The February 1863 publication of Lyell's

1. Darwinian evolution is the only theory from this period whose reception has been examined in detail. The literature is extensive, and the following list is meant to be suggestive rather than exhaustive. *Darwin and His Critics,* ed. D. L. Hull (1973) offers a valuable collection of contemporary scientists' commentaries on *Origin of Species* and contains a useful introduction. A. Ellegard, *Darwin and the General Reader* (1958) surveys Darwin's reception in the British popular press. *The Comparative Reception of Darwinism,* ed. T. F. Glick (1974), places the British response to Darwin in international perspective. J. R. Moore, *The Post-Darwinian Controversies* (1982) and J. Lyon, "Immediate Reactions to Darwin," *Church History* 41 (1972): 78–93, discuss the reactions of the Protestant and Catholic press.

2. R. M. Young, "Natural Theology, Victorian Periodicals, and the Fragmentation of the Common Context," in *Darwin's Metaphor* (1985).

widely popular *Antiquity of Man* opened the literary floodgates.[3] Every major general-interest review—both secular and religious—reviewed Lyell's book, and scores of additional articles explored its implications. Commentaries on the new case for human antiquity also appeared in books and articles discussing a variety of related issues: evolution, archaeology, anthropology, and the relationship between science and religion.

The generalizations in this chapter are based on eighty-six books and articles, published in Britain between 1859 and 1875, that discuss the new case for human antiquity formulated in 1858–63 (see App. 2). These eighty-six items include articles that appeared in major scientific, religious, and general-interest periodicals, as well as books that treated the subject either fully or in passing. The sample excludes works by the five men whose investigations established the new case for human antiquity: Pengelly, Falconer, Prestwich, Evans, and Lyell. Their writings on the human antiquity problem after 1863 consisted of reports of new sites rather than commentaries on particular theories. It also excludes the major works of John Lubbock, William Boyd Dawkins, and other founders of prehistoric archaeology. These works, though chronologically within the bounds of the survey, are intellectually distinct from the other commentaries. Chapter 7 considers them and the new subdiscipline for which they represented declarations of independence.

The eighty-six items discussed here do not exhaust the material available, but they fairly represent the range of opinions on human antiquity held by educated Britons. They are the work of scientists, clergymen, and lay writers from a variety of backgrounds. Roughly one-third of the commentaries, mostly those appearing in religious journals and the leading general-interest reviews, were published anonymously. Thirteen remain anonymous, and the identities of two other authors remain hidden behind impenetrable pseudonyms. Allowing for some duplication of authorship among the anonymous commentaries, the eighty-

3. Lyell stated, in a letter written shortly after the appearance of the second edition, that the first edition of 4,000 copies had sold out in nine weeks and that the fifth thousand was also selling briskly. (Lyell to George Poulett Scrope, April 29, 1863, Lyell Papers, American Philosophical Society [APS], B L981.16).

six items in my survey probably represent the work of sixty to sixty-five different authors.

The issues discussed in the debate over human antiquity varied with time. The earliest commentaries focused on the validity of the new case for human antiquity, and on the evidence for men among the mammoths. By 1863, most writers' concerns had changed. The *age* of the first humans, rather than their presence among now-extinct animals, became the crucial issue. The reasons for this change, the wide range of opinions expressed on the age of the human race, and the ways in which the new case for human antiquity became a buttress for—rather than a challenge to—existing ideas about the early human race all reflect the sense of intellectual crisis that prevailed in mid-Victorian Britain.

The Structure of the Debate

The October 1860 issue of *Blackwood's Edinburgh Magazine*, featured an article by expatriate American geologist Henry Darwin Rogers, "The Reputed Traces of Primeval Man." Rogers had visited the Somme Valley earlier in the year, and his article was the first to deal at length with the discoveries made there by Prestwich, Evans, and Boucher de Perthes. The article identified, with great clarity and precision, the questions that lay at the core of the debate over human antiquity. (1) Were the chipped flints human artifacts? (2) Did the flints' stratigraphic position imply the coexistence of humans and extinct animals? and (3) How old were the bones among which the flints were found? Every published commentary on the new case for human antiquity attempted to answer one or more of the three questions; most treated more than one, and a significant number grappled with all three. The attention paid to each of the three changed significantly over time, however. As a result, it is possible to divide the debate into two distinct phases, separated by the publication of Lyell's *Antiquity of Man* in February 1863.

The first phase of the debate began in the summer of 1859, only weeks after Joseph Prestwich and John Evans had announced the results of their April visit to the Somme Valley. It focused almost exclusively on the first two of Rogers's three questions, the man-made status of the chipped flints and their age relative to the bones of extinct animals. Only three of the twenty-four items appearing between May 1859 and January 1863 took a clear position on the third question, the age of the remains of extinct animals. The rest either remained silent on the question or argued that it could not be solved until more data, or different types of data, became available.

Nineteen of the twenty-two authors who took part in the first phase of the debate were scientists, and their commentaries were exclusively concerned with the scientific validity of the new case for human antiquity. Few discussed its potential impact on established scientific ideas, and fewer still assessed the religious implications of human antiquity. The new case was criticized primarily by those outside the geological community, including archaeologist Gardner Wilkinson, botanist John Henslow, physician Henry Ogden, and several anonymous writers in religious periodicals.[4] Its principal defenders were geologists, including Joseph Beete Jukes, David Ansted, Richard Owen, Roderick Murchison, and Henry Darwin Rogers.[5] The arguments deployed by both critics and defenders of the new case were little more than expansions on those used in summer 1859, when Prestwich and his colleagues defended their work against attacks by Thomas Wright and John Anderson.

The growing stock of corroborative data collected in the early 1860s gradually changed the terms of the debate. The discovery of greater numbers of finely chipped flint objects gradually rendered untenable the argument that the flints had been formed by natural forces. The careful documentation of an increasing number of sites where stone tools had been found among the bones of extinct animals diminished the chances that all such associations were accidents. A year-by-year analysis of the first phase of

4. G. Wilkinson, "Remains of Man in Caves," *Athenaeum*, June 9, 1860, 791. J. S. Henslow, "Flint Weapons in the Drift," *Athenaeum* February 11, 1860, 206–7. H. Ogden, "The Flint Find," *Athenaeum*, November 5, 1859, 666. "Review of *Geological Gossip*," *British Quarterly Review* 32 (1860): 258–59.

5. J. B. Jukes, *Student's Manual of Geology*, 2d ed. (1862), 700–703; D. T. Ansted, "The Antiquity of Man," in *Geological Gossip* (1860); R. Owen, *Paleontology* (1860), 401–3; R. I. Murchison, "Thirty Years Retrospect," *American Journal of Science* 83 (1862): 1–2; H. D. Rogers, "Reputed Traces of Primeval Man," *Blackwood's* 88 (1860): 422–38.

the debate clearly shows the effect of the geologists' accumulating stock of data. Of the ten criticisms of the new case for human antiquity that appeared between May 1859 and January 1863, seven appeared in 1859 and two in 1860. Attacks on the authenticity of the stone tools or the coexistence of humans and extinct animals had virtually ceased by the beginning of 1861.

This shift in attitudes was also apparent on an individual level. After arguing vehemently in the *Athenaeum* that the chipped flints of the Somme Valley were natural, Thomas Wright publicly accepted their man-made status in 1861. John Henslow, the respected Cambridge botanist, underwent a similar conversion. Henslow's first commentary on human antiquity criticized the data that Joseph Prestwich had gathered at Hoxne. His later writings—published, like the first, as letters to the *Athenaeum*—accepted the data but, argued that they did not prove the coexistence of humans and extinct animals. Henslow emphasized, in a letter written after he returned from a visit to the Somme Valley, that the basic issue in discussions of human antiquity was what constituted adequate proof. "The facts I have witnessed do not *of necessity* support the hypothesis of a prehistoric antiquity for these works of man. Neither do I consider that the bones of extinct animals found associated with them must *of necessity* be supposed to have belonged to individuals cotemporary with the uncunning workmen who wrought the rude hatchets." Henslow took a similar position in a February 1861 letter to a Mr. Nihill and in a public lecture delivered at Ipswich in the same month.[6] Fellow botanist Joseph Hooker believed, however, that Henslow had finally accepted the evidence for men among the mammoths and was preparing a fresh lecture on the subject when he died in May 1861.[7]

Charles Lyell's *Antiquity of Man* summarized and organized the corroborative data gathered in the 4½ years after Prestwich and Evans's trip to Abbeville. It virtually ended debate over the authenticity of the stone tools from the drift and their makers'

6. J. S. Henslow, "Flints in the Drift," *Athenaeum*, October 20, 1865, 516; Henslow to "My Dear Nihill," February 14, 1861, APS, B H382; notice of the Ipswich lecture appeared in the *Edinburgh New Philosophical Journal* 14 (1861): 171.

7. L. Jenyns, *Memoir of the Reverend John Stevens Henslow* (1862), 216.

coexistence with extinct animals. All but the most determined opponents of the new case conceded those two points, while supporters cited Lyell rather than undertaking their own detailed defense. Only five of the roughly sixty commentaries that appeared between 1863 and 1875 questioned the authenticity of the tools or their contemporaneity with Post-Pliocene mammals. Four of those five, moreover, were written by a single heir to Thomas Wright's discarded criticisms, civil engineer Nicholas Whitley. The response to Whitley's interpretation of the chipped stone tools as accidents of nature was, even among sympathetic listeners, skeptical at best.[8]

Antiquity of Man changed the nature of the human antiquity debate in other ways as well. Most important, it made the third of Rogers's three questions the center of all subsequent discussions of the age of the human race. Until 1863, commentators on the new case for human antiquity had been concerned with the presence (or nonpresence) of men among the mammoths. After Lyell's book, and as a result of it, they turned their attention to the age of both men and mammoths.

The key to understanding the debate over human antiquity after 1863 is the distinction between relative and absolute age. Statements about the relative age of an object—a fossil bone or stone tool, for example—take the form, "Object X is older than object A but not as old as object B." Statements about the absolute age of the same object take a different form, "Object X is n years old." Early- and mid-Victorian geologists routinely determined the relative ages of strata and their fossil contents. The methods they used were straightforward and the results were, with a few celebrated exceptions, seldom controversial. Determining absolute ages was, in contrast, a methodological nightmare. Until the discovery of radioactive decay in the twentieth century, geologists had no means of measuring the absolute ages

8. Nicholas Whitley's commentaries on human antiquity include *Are the Flint Implements from the Drift Authentic?* (1865), "True and False Flint Weapons," *Popular Science Review* 8 (1869): 30–38, "The Paleolithic Age Examined," *Journal of the Transactions of the Victoria Institute* 8 (1874–75): 3–23 (hereafter *J. Trans. Vict. Inst.*), and "Brixham Cave and Its Testimony to the Antiquity of Man Examined," in ibid. 211–24. For contemporary reactions to Whitley's ideas, see the record of the discussion that followed "The Paleolithic Age Examined," in ibid., 23–50.

of rocks. Nineteenth-century calculations of absolute age were, therefore, no more than estimates—estimates based on highly controversial assumptions about the tempo of geologic change.

The stratigraphic data gathered by Prestwich and Evans in the Somme Valley made the relative age of the stone tools they collected virtually self-evident. The tools had been found in undisturbed layers of gravel that Prestwich's earlier stratigraphic studies showed to be part of the Drift. They all came from deposits that lay above the "Boulder Clay," a stratum associated with the glacial period of the early Post-Pliocene epoch. Finally, the tools lay among the bones of extinct animals, such as mammoths, that had been common in Western Europe after the end of the glacial period. The three lines of evidence pointed to a single conclusion: the toolmakers of the Somme had lived after the glacial period, but before the extinction of the mammoth. This placed them firmly in the latter half of the Post-Pliocene.

Calculating the absolute age of the toolmakers and the extinct animals that they hunted was problematic. The stone tools found at Menchecourt, for example, lay twenty-five to thirty feet below the surface of the ground, covered by beds of gravel, loam, and clay. The stratigraphy of the area made it clear that the overlying beds had been laid down before the valley of the Somme had been eroded to its modern width and depth. It was equally clear that even the youngest of the beds overlying the tools was older than the twenty to thirty feet of peat that formed the valley floor.[9] Determining the absolute age of the tools meant figuring the time necessary for (1) the deposition of the overlying gravel and loam; (2) the erosion of the valley to its present contours; and (3) the formation of as much as thirty feet of peat on the valley floor. On that much, British geologists agreed. Their opinions on what figures should be used in the calculations—and whether such calculations were even worthwhile—varied widely.

This variation was evident even among the geologists who had formulated the new case for human antiquity. Joseph Prest-

9. C. Lyell, *Geological Evidences of the Antiquity of Man* (1863), 122; J. Prestwich, "Theoretical Considerations," *Philosophical Transactions of the Royal Society* 154 (1864): 256–64.

wich and Charles Lyell, two of the leading figures in the investigation, differed sharply and publicly over the absolute age of the Post-Pliocene toolmakers. Lyell, committed to uniformitarianism, argued that geological processes such as erosion and deposition had proceeded gradually in the Post-Pliocene, just as they do in the present. He concluded that the volume of the gravel and loam beds, the roundness of chalk and flint pebbles imbedded in them, and the extensive widening and deepening of the valley all bespoke "a vast lapse of time." Lyell also concluded, on the basis of fossil evidence, that the "thousands of years" represented by the peat was a small fraction of the time separating the tool-bearing gravel from the present. *Antiquity of Man* thus implied that humans had apparently lived in Europe close to 100,000 years ago. A year later, addressing the British Association, Lyell made his claim of a thousand centuries explicit.[10]

Prestwich rejected Lyell's assumption about the uniformity of geological processes as gratuitous. A careful study of the geological facts, Prestwich argued, showed that the Somme Valley had been shaped not only by the gradually acting forces that now held sway, but also by periods of intense flooding. These floods had, he believed, excavated the valley and laid down much of the gravel found along its banks. They had done so, moreover, in significantly less time than Lyell assumed was necessary to produce the same results. The toolmakers and the now-extinct mammals that they hunted had, Prestwich concluded, probably lived about 20,000 years in the past.[11]

This difference of opinion became public between 1862 and 1864, aired in two papers by Prestwich and in the pages of *Antiquity of Man*. For both Lyell and Prestwich, it was simply one aspect of the geological community's ongoing work on the age of the human race. Though each argued explicitly for his view, neither devoted much space to the dispute in his published work. Outside the human antiquity core set, however, the situation was very

10. Lyell, *Antiquity of Man*, 144–46; C. Lyell, "Presidential Address," *BAAS Report* 34 (1864): lxxiv.

11. The most detailed statement of this case is Prestwich's "Theoretical Considerations," particularly pp. 264–66 and 298–303. Though published in 1864, the paper is actually a composite of two shorter papers written in 1862.

different. Commentaries on human antiquity written before 1863 treated the absolute age of the human race as a relatively minor issue. Commentaries written after mid-1863, however, treated it as virtually the only significant issue. The publication of Lyell's book turned the debate over human antiquity into a general referendum on the question of absolute age.

The emergence of absolute age as a dominant issue changed the human antiquity debate in other ways, as well. Until 1863, most of the commentaries that appeared in print (81 percent) were written by scientists. During 1863, clergymen and lay people from a variety of backgrounds began to take an active role. The tone of the debate also changed, as heated and visibly partisan commentaries began to appear for the first time. Some writers—the Duke of Argyll, for example—continued to insist that there was not sufficient data to resolve the absolute age question, but the popularity of such a position steadily diminished after 1863.

There were a variety of causes for this radical transformation, some internal to the debate and some external. The most prominent was *Antiquity of Man* itself. Lyell's use of uniformitarian assumptions to calculate the tools' absolute age tied the debate over human antiquity to a much older debate over geological methods. Lyell's uniformitarianism was more than thirty years old when *Antiquity of Man* appeared, and it had been controversial for most of that time. A succession of critics argued, as Prestwich had, that it was unreasonable and unscientific to assume that the intensity of geological processes had never substantially changed. To do so, they contended, was to allow preconceptions to distort one's vision of the facts, to see nature through the lens of theory. The very magnitude of Lyell's estimate of absolute age contributed to his critics' belief that it was a product of faulty reasoning. To argue that the human race was older than once suspected was one thing; to increase its age tenfold at a single stroke was another.

Lyell and his defenders, Darwin among them, responded that assumptions about the congruity of past and present geological processes were all that made a scientific understanding of the past possible. Lyell also defended the magnitude of his estimate.

Speaking as president of the British Association for the Advancement of Science in 1864, he told of a "great Irish orator" who, born in poverty, remained frugal even after becoming wealthy. The same habitual frugality, Lyell argued, had shaped scientists' views of the past. "Throughout our early education we have been accustomed to such strict economy in all that relates to the chronology of the earth and its inhabitants in remote ages, so fettered have we been by old traditional beliefs, that even when our reason is convinced, and we are persuaded that we ought to make more liberal grants of time to the geologist, we feel how hard it is to get the chill of poverty out of our bones."[12] To follow Lyell in defending a 100,000-year figure for humankind's absolute age was, therefore, to take a methodological position as well as a chronological one.

The conclusion that humans had existed in the late Post-Pliocene had comparatively little impact on scientific or religious thought so long as the age of the late Post-Pliocene remained undefined. The absolute age of the human race, on the other hand, had potentially enormous implications for both science and religion. The emergence of absolute age as a major issue therefore made the human antiquity debate important in intellectual circles far outside the geological community. As an example, the idea that humans had lived among mammoths and other now-extinct beasts posed little threat to traditional interpretations of the Bible. The idea that those humans had lived 20,000—much less 100,000—years ago threatened to completely overturn the popular understanding of the early chapters of Genesis.

The timing of the new case for human antiquity was, in this particular case, as significant as its intellectual content. The publication of the multiauthored volume *Essays and Reviews* in 1861 and J. W. Colenso's *The Pentateuch and the Book of Joshua Critically Examined* in 1862 created an air of crisis in the British church that cut across denominational boundaries. Both works were seen as attacks on the divinely inspired status of the Bible. The appearance of Darwin's *Origin of Species* and of subsequent works that applied its principles to humans contributed to the sense of crisis

12. Lyell, "Presidential Address," lxxiv.

by challenging traditional understandings of God's relationship to the natural world. The absolute age of the human race was— or could be—tied closely to both the hermeneutic and biological challenges to orthodoxy. The intellectual crisis of the early 1860s not only promoted interest in the human antiquity debate but also raised significantly the stakes of the debate.

The Problem of Absolute Age

The debate over the absolute age of the human race involved a wide spectrum of views. The most conservative participants embraced the 6,000-year figure calculated by such seventeenth-century biblical scholars as James Ussher and John Lightfoot. The most liberal went far beyond the 100,000 years estimated by Lyell in *Antiquity of Man*. Between these extremes lay a variety of intermediate positions, of which Joseph Prestwich's estimate of 20,000 years was the most popular.

Not all commentators on the problem of absolute age expressed their views precisely. A few discussed the problem only in theoretical terms, declining, as Rogers and Argyll had before 1863, to draw a specific conclusion. Others, particularly nonscientists, took specific positions but expressed them only in qualitative terms: "a long series of ages" or "very great antiquity." It is nonetheless possible to attach a specific estimate of absolute age to most of the commentaries written in 1863 or later, either directly or by inference from the author's views on uniformitarianism, the accuracy of Genesis, or similar issues. The majority of these estimates fall into three distinct camps that can be labeled Lyellian (100,000 years or more), Prestwichian (ca. 20,000 years), and Traditional (6,000–8,000 years).

Each of the three camps attracted some scientists, some religious writers, and some members of the general public. Religious writers outnumbered scientists in the Traditional camp, and the opposite was true in the Lyellian camp, but there was no clear-cut distinction between a "scientific" and a "religious" position in the debate over absolute age. Because the debate turned not on facts but on the interpretation of facts, allegiance to a particular camp frequently reflected a participant's prior intellectual commitments—theoretical, methodological, or religious. The most profitable way to survey the texture of the debate and the impact of these various influences is to examine each of these camps in turn.

THE LYELLIANS. Lyell's premise that geologic change takes place in tiny increments over vast spans of time was the key to his interpretation of Earth history. It also lay at the heart of his work on the absolute age of the human race. The 100,000-year figure for human antiquity that Lyell proposed in 1863 was a direct product of his belief in gradual change. Though many members of the Lyellian camp declared 100,000 years to be too conservative, none questioned the reasoning behind it.

The first writer to give the human race an absolute age of Lyellian proportions was Lord Wrottesley, an astronomer who served as president of the BAAS in 1860. Wrottesley's presidential address discussed the still-fresh discoveries in the Somme Valley and commented favorably on Lyell's address to the BAAS geological section the previous year. He concluded that "a great lapse of time" separated the Somme Valley toolmakers from the present day and offered a detailed summary of Lyell's arguments in support.[13] Lyell's claim, in a letter to archaeologist Albert Way, that he had ghost-written the relevant section of Wrottesley's address does not diminish the address's significance.[14] Wrottesley remains the only commentator to publicly support Lyell's views before the 1863 publication of *Antiquity of Man.*

The rest of the commentaries that make up the Lyellian camp began to appear in 1863. The most prominent and most enthusiastic were written by leading scientists concerned with the anthropological implications of human antiquity. They embraced, and in many cases expanded, Lyell's estimate of the human race's absolute age. More important, many took a step that Lyell had been unwilling to take: they introduced the issue of human antiquity into the debate over whether Darwin's theory of evolution applied to the humankind as well as the lower animals.[15]

T. H. Huxley's *Man's Place in Nature* was not seen, at the time

13. Lord Wrottesley, "Presidential Address," *BAAS Report* 30 (1860): lxii–lxiii.
14. Lyell to Albert Way, August 9, 1860, Lyell Papers, APS, B L981.18.
15. M. J. Bartholomew, "Lyell and Evolution," *BJHS* 6 (1973): 261–303.

of its publication in 1863, as the revolutionary work that its modern reputation makes it out to be.[16] It contained little of the rhetorical brilliance that became Huxley's trademark and no impassioned defense of Darwin's theory or its applicability to humans. Suggestive rather than explicit, *Man's Place in Nature* was nonetheless a major contribution to the emerging debate over human evolution. The book consisted of three separate essays linked by an implicit common theme: the relationship between humans and apes. The first essay surveyed the anatomy of the great apes (chimpanzees, gorillas, and orangutans); the second compared their anatomy to that of modern humans; the third discussed the oldest known human fossils and compared them to modern humans. All three essays were primarily catalogs of data. Read with Darwin's theory in mind, however, the second essay in particular constituted a strong case for the descent of humans and apes from a common ancestor. Huxley made the same point explicitly, albeit briefly, in the last essay.

This third essay, taken at face value, was a cautious anatomical analysis of Schmerling's Engis skull and the bones of the Neanderthal Man. Its thesis was clear. Neither the Engis skull nor, for all its "apish" appearance, the Neanderthal skull possesses a brain capacity outside the range exhibited by modern humans. Neither, therefore, is even an approximate representation of "that lower pithecoid form, by the modification of which [man] has, probably, become what he is."[17] Those who argued against the evolutionary origins of the human race rejoiced in this conclusion.[18] Huxley, however, took a different approach to it. If the oldest known human fossils took us no closer to the common ancestor of humans and apes, he argued, then we must seek the remains of that ancestor in still older deposits.

16. T. H. Huxley, *Man's Place in Nature and other Anthropological Essays* (1899). The 1899 edition was issued as part of Appleton's uniform edition of Huxley's works. It contains the three essays from the 1863 edition as well as three additional, later essays. For an analysis of *Man's Place in Nature* and its contemporary reception see M. Di Gregorio, *T. H. Huxley's Place in Natural Science* (1984), 139–59.

17. Huxley, "On Some Fossil Remains of Man," in *Man's Place,* 186–206, 209, quotation on 209.

18. J. Crawfurd, "Notes on Sir Charles Lyell's *Antiquity of Man,*" *Anthropology Review* 1 (1863): 172–73; Duke of Argyll, *Primeval Man* (1870), 70–75; J. W. Dawson, "Primitive Man and Revelation," *J. Trans. Vict. Inst.* 8 (1874–75): 53–55.

Early in his essay on human fossils, Huxley cited evidence gathered by Lyell to show that the Engis skull was of Post-Pliocene age. To the Neanderthal skeleton he assigned a "great, though uncertain, antiquity."[19] "Older deposits," therefore, meant strata laid down in the Pliocene, or perhaps even the Miocene epoch. Making this point in the final paragraph of *Man's Place in Nature*, Huxley underlined what he saw as the intimate connection between the human race's age and its origin. "[If] any form of the doctrine of progressive development is correct," Huxley concluded, "we must extend by long epochs the most liberal estimate that has yet been made of the antiquity of man."[20]

Though Huxley did not refer to Lyell's calculations specifically, his readers would certainly have made the connection. John Lubbock's review of *Antiquity of Man*, published in the *Natural History Review* in April 1863, stated the connection plainly. Lubbock later achieved international fame for his work in anthropology and prehistoric archaeology, but in 1863 he was known primarily as a biologist. His review contended that geologists were not the only scientists whose work had been instrumental in establishing human antiquity. Though geologists have apparently led the way, he concluded, "the Zoologist is really in advance, and claims for man a higher antiquity than even Sir Charles Lyell can at present bring himself to admit." Lubbock's reasoning recapitulated Huxley's. Evolutionary theory demands the presence of human ancestors in times much older than the oldest human fossils.[21]

Most scientists in the Lyellian camp used variations on Huxley's argument. Alfred Russel Wallace, though he denied that natural selection was the sole cause of human evolution, agreed that evolutionary theory implied the presence of humans in the Pliocene or Miocene epoch. He regarded 100,000 years as a minimum estimate of humankind's time on Earth and suggested that the actual figure might be a million years or more. David Page and J. F. McLennan substituted cultural for biological development but drew similar conclusions. The chipped-stone tools from

19. Huxley, "Fossil Remains," 158.
20. Ibid., 209.
21. [J. Lubbock], "Review of *Antiquity of Man*," *Natural History Review* 3 (1863): 212, 217, quotation on 212.

the Drift were products of a savage culture, they argued; humankind's gradual rise to mental, moral, and technological sophistication must have taken hundreds of centuries.[22]

These scientists' views on the absolute age question were formed principally by their commitment to theories of biological or cultural evolution. All four cited the implications of evolutionary theory rather than Lyell's geological calculations as grounds for accepting or extending the 100,000-year figure. Several of them, moreover, sharply criticized *Antiquity of Man* as a highly derivative work hardly worthy of its author. Darwin's disappointment with Lyell's tepid defense of evolution is well documented, and Lubbock's scorn for Lyell's conservatism is evident in his review. The evolutionists' belief that humans were as old as Lyell claimed reflected the concurrence of their research with his rather than an endorsement of his work.

Roughly half of Lyell's support came from leading evolutionists and took the form of books and scientific papers. The balance of the pro-Lyell commentaries were written by nonscientists and appeared in general-interest periodicals. The commentators in the second group took a very different approach to their subject than did the evolutionists. They focused less, for example, on the scientific than on the religious implications of the human race's great absolute age. More important, they accepted uncritically Lyell's geological arguments for a high absolute age and restated those arguments at some length in their commentaries. They were, in fact, the only group of commentators to accept Lyell's assumptions and calculations at face value.

"Sir Charles Lyell's volume is an elaborate assault on the popular chronology," wrote an anonymous reviewer in the *Athenaeum*, "either the scientific or the popular chronology must be wrong." The difference between the two, he continued, might well be thousands of centuries.[23] The *Athenaeum*'s reviewer ele-

22. A. R. Wallace, "The Origin of the Human Races and the Antiquity of Man Deduced from the Theory of Natural Selection," *Anthropology Review* 2 (1864): clviii–clxx. For Wallace's ideas on human evolution, see H. L. McKinney, *Wallace and Natural Selection* (1972); and M. Fichman, *Alfred Russel Wallace* (1981), chap. 4. D. Page, *Man: Where, Whence, and Whither?* (1868), 116–35; [J. F. McLennan], "The Early History of Man," *North British Review* 50 (1869): 272–90.

23. "Review of *Geological Evidences of the Antiquity of Man*," *Athenaeum*, February 14, 1863, 219.

gantly stated a theme that ran through all the pro-Lyell commentaries written by nonscientists. The reviewer implied that Lyell's view and the "scientific" view were identical: that Lyell spoke not just for himself, but for the whole geological community. "'Time, time, and yet more time' is the cry of the student of antiquity," exclaimed the *Saturday Review,* placing all geologists and a variety of other scientists in the Lyellian camp. The *Westminster Review* concluded that "all the facts hitherto brought to light" unanimously suggest that a "long series of ages" lay between modern humans and the toolmakers of the Drift.[24] Those who argued for a low absolute age, it implied, ignored the facts and violated the principles of solid Baconian science.

The roots of this enthusiasm for Lyell's work are not difficult to locate. His *Principles of Geology, Elements of Geology,* and addresses to the BAAS had kept him and his uniformitarian ideas in the public eye for more than thirty years. He was Britain's best known—and, from the nongeologist's point of view, most prolific—geological theorist. It is hardly surprising that those outside the geological community, not privy to the theoretical disputes taking place behind closed doors at the Geological Society, equated Lyell's views with those of his lesser-known colleagues. The unanimity that these nonscientist commentators envisioned was illusory, however—a fact that several of Lyell's colleagues discussed at length in their own reviews of *Antiquity of Man.*

THE PRESTWICHIANS. Joseph Prestwich's first paper on the stone tools found in the Somme Valley laid out his position on their absolute age. The geological evidence, he told the Royal Society in May 1859, showed that the tools were older than ordinary chronologies allowed. Prestwich quantified this position in later papers, settling on an age of roughly 20,000 years and so claiming a middle ground between the Lyellians and the traditionalists. The commentators who joined Prestwich on this middle ground formed the smallest and most uniform of the three groups discussed here. Nearly all were scientists, and all shared Prestwich's conservative views on proper scientific method. Their writings,

24. "Review of *Geological Evidences of the Antiquity of Man,*" *Saturday Review,* March 7, 1863, 312; "The Antiquity of Man," *Westminster Review,* 79 (1863): 290.

among the most prominent in the debate over absolute age, counseled caution and sobriety in dealing with the still fragmentary evidence of Post-Pliocene humans.

The Prestwichians, unlike the most prominent Lyellians, often used geological arguments to make their case. Though the structure of these arguments varied, the theme remained constant: the strata overlying the stone tools and mammoth bones of the Somme Valley had not accumulated gradually and so did not represent an immense span of time. John Phillips and James Forbes, both leading geologists, each discussed the absolute age problem on several occasions. Their articles—even those in general-interest periodicals—detailed the geology of the Somme Valley and catalogued features that were inconsistent with Lyell's vision of gradual change. They pointed, for example, to multi-ton sandstone blocks imbedded in the gravel—blocks that present-day floods could not have shifted.

Phillips and Forbes also deployed a second type of geological argument, noting that the coexistence of humans and mammoths proved nothing about the absolute age of either. Such a coexistence offered no more ground, Phillips observed, "for admitting man to be of immense antiquity, than for allowing to the huge pachyderm the advantage of living nearer our own time."[25] This claim, repeated by every member of the Prestwichian camp, may have been designed to refute assumptions that absolute and relative age must be closely related. It carried, therefore, a subtle message about what the Prestwichians saw as proper scientific methods: check your preconceptions at the door.

The Prestwichians' concern with methodology was central to their criticisms of Lyell's solution to the absolute age question. His calculations, they believed, rested on too many a priori assumptions and too little solid data. They regarded Lyell's use of modern geological processes as models for ancient ones as a particularly egregious error. Forbes concluded one scathing assessment by deeming it "an abuse of logic and of the rules of evidence."[26] Phillips, a critic of Lyellian uniformitarianism since the

25. [J. Phillips], "Review of *Geological Evidences of the Antiquity of Man*," *Quarterly Review* 114 (1863): 417.

26. [J. D. Forbes], "Lyell on the Antiquity of Man," *Edinburgh Review* 118 (1863): 292.

1830s, shared Forbes's doubts about the equivalence of past and present processes and his belief that the available evidence for human antiquity did not support Lyell's immense claims. "There is certainly no warrant," he concluded, "for proceeding many steps in this direction, along a slippery path, over which time has gathered many shadows, and along which the torch of science sheds but a feeble and unsteady light."[27]

Phillips and Forbes were the most influential, but not the only, members of the Prestwichian camp. John Henslow's conversion to a firm belief in human antiquity placed him on the same middle ground. John R. Young, professor of mathematics at the University of Dublin, defended Prestwich's approach to the absolute age question in an 1865 book on a more general question, entitled *Modern Scepticism in Relation to Modern Science*. Young joined Phillips and Forbes in criticizing Lyell's methods, but his charges were more extreme and his tone more strident. Scientists, he argued, should follow the example set by Newton. They should work to formulate laws that describe the physical world but be willing to admit their ignorance of causes. Young believed that "speculative geologists" routinely violated this principle by assuming the equivalence of past and present geological forces. They had, moreover, compounded their error by ignoring the same assumption when it conflicted with some new speculation.[28]

Those scientists who adopted Lyell's position on absolute age saw themselves as advocates for a specific theory. Those who followed Prestwich, on the other hand, saw themselves primarily as defenders of "good science." Though they presented a particular interpretation of the data, their intellectual commitment was less to the interpretation than to the integrity of these data. They opposed Lyell both because they found his views on absolute age implausible, and because they regarded those views as the product of sloppy reasoning.

The Prestwichians were, in this regard, closely allied with the

27. [Phillips], "Review of *Antiquity of Man*," 417. For Phillips's early opposition to Lyell, see J. Morrell, "Science and Government," in *Science, Politics, and the Public Good*, ed. N. Rupke (1988), 14.

28. J. R. Young, *Modern Scepticism in Relation to Modern Science* (1865), 143–58. Young went on to raise the same charges against Darwin and Huxley, 158–74.

many commentators who abstained from the debate over absolute age. Most who declined to draw conclusions about humankind's absolute age justified their decision by claiming insufficient evidence. The most scientifically sophisticated among them—H. D. Rogers and the Duke of Argyll, for example—noted that the absolute age question was but one facet of a larger problem: whether past and present geological forces are (as Lyell assumed) equivalent. Until the larger issue was settled, the smaller one could not be.[29] Writing in 1874, a young geologist named William Boyd Dawkins suggested that the absolute age question was insoluble. Attempts to quantify the age of the human race were as fanciful as Ussher's calculations of the date of the Creation.[30]

A majority of those who abstained, however, simply counseled caution and sobriety. "While facts are every day multiplying upon us, much is yet needed for that thorough confirmation which science requires," wrote Henry Holland in 1864, and many agreed with him.[31] The lack of solid data on rates of geologic change meant that nineteenth-century writers could defend the wait-and-see position indefinitely. A decade after Holland, Rev. George Deane implied that little had changed, and that the time for erecting a theory had not yet arrived. "The subject," he told readers of the British Quarterly Review, "is one on which we can afford to wait; weighing and sifting carefully, meanwhile, the accumulating evidence."[32]

Refusal to draw conclusions from a limited set of facts was,

29. Rogers's correspondence makes clear his opposition to Lyell's strict uniformitarianism and belief in occasional "paroxysmal actions." See, e.g., H. D. Rogers to W. B. Rogers, December 23, 1859, in E. Rogers, The Life and Letters of William Barton Rogers (1896), 2:17–18. Rogers apparently did not discuss his views on the tempo of geologic change in his 1860 article because he intended the article to be instructional rather than partisan. Letters that passed between Rogers and his brother William during September 1860 (reprinted in E. Rogers, Life and Letters, 2:40–44), discuss the purpose and content of the article.

30. [W. B. Dawkins], "The Antiquity of Man," British Quarterly Review 58 (1874): 352–55.

31. [H. Holland], "Man and Nature," Edinburgh Review, American ed. (1864): 243.

32. G. Deane, "Modern Scientific Inquiry and Religious Thought," British Quarterly Review 59 (1874): 52.

according to the principles of Baconian scientific method, not merely acceptable but virtuous. To take such a course was implicitly to criticize those who, when faced with the same limited set of facts, quickly founded a theory upon them. Strict Baconianism may have begun to lose its stranglehold on British science by the 1860s, but participants in scientific debates could still seize the high ground by claiming that restraint in theorizing was the "philosophical" course. Those who abstained from the absolute age debate may have meant their actions to be a criticism of Lyell—to imply that acknowledging the data's limits was more prudent than wildly speculating on their meaning.

THE TRADITIONALISTS. The traditional view of human antiquity, which assigned the human race an absolute age of little more than 6,000 years, remained popular in Britain well into the 1870s. The scientific community had virtually abandoned it by 1863, but it enjoyed broad support among clergymen and devout laymen. The popularity of the traditional view was due, in part, to its roots in and harmony with traditional interpretations of Genesis. The belief that it should be defended was strengthened by a sense that the foundations of Christian belief—particularly the divine inspiration of Scripture—were under attack. Defenders of the traditional view did more, however, than simply reassert biblical authority. Their arguments were scientific rather than scriptural and reflected a detailed knowledge of the arguments aired in secular books and periodicals.

Commentators who estimated the human race's absolute age at 6,000–8,000 years generally used one of three principal strategies. The most audacious was to argue that humans had appeared, and the great post-Pliocene mammals had become extinct, little more than 6,000 years ago. Canadian geologist J. W. Dawson, the only scientist to publicly defend the Traditional view, developed this theory at length in *The Story of the Earth and Man* and many subsequent works.[33] The challenge in defending such a theory was to explain the significant geological changes

---

33. J. W. Dawson, *The Story of the Earth and Man* (1873). For a clergyman's variation on the same theme, see An Essex Rector, *Man's Age in the World* (1865).

that had taken place in Western Europe since the end of the Post-
Pliocene. Dawson himself invoked a "comparatively rapid sub-
mergence and re-elevation of the land" which took place soon
after the advent of man.[34]

Dawson's argument depended on its author's detailed familiarity with geological processes and the history of life on Earth.
The idea that changes for which Prestwich allotted 20,000 years
might have taken only 6,000 strained most geologists' credulity,
but Dawson's ability to marshal a wide range of supporting evidence reduced the strain. A prolific amateur interpreter of scripture, Dawson was also capable of reconciling his science with his
religion.[35] Most advocates of the Traditional view lacked Dawson's scientific expertise, however, and so took a different approach.

The second, most popular, defense of the Traditional view
appealed, as did the Prestwichians, to Baconian methods. It regarded all attempts to extend the absolute age of the human race
much beyond 6,000 years as hypotheses that were, at best, not
clearly substantiated by the available facts. The essence of this
defense was its appeal to scientific caution. Those who employed
it argued that, since the facts were sparse and their meaning uncertain, the "philosophical" course of action was to reserve all
theoretical judgment. Reserving judgment on humankind's absolute age explicitly meant, of course, leaving the traditional figure
of 6,000 years in place until the accumulation of additional facts
could settle the issue. This was defense by default, and while it
lacked the forcefulness of Dawson's work it also avoided the appearance of being special pleading for a Bible-based view of
nature.

These appeals for caution were different from those raised by
the Prestwichians and those who abstained from the absolute age
controversy. Members of the latter two groups agreed that the
human race was older than once thought, arguing only that calls

34. Dawson, *Earth and Man*, 290.
35. J. W. Dawson, *Archaia* (2d ed., 1860) is both a detailed example of Dawson's attempts to harmonize science and Scripture and a defense (against the
claims of professional biblical scholars) of his right to interpret Scripture for
himself.

for vast spans of time were suspect. Traditionalists who appealed for caution argued that *any* significant expansion of the old 6,000-year estimate was unjustified. The extent to which their appeals reflected genuine concern with method, as opposed to rhetorical convenience, is an open question that can, for the purposes of this discussion, remain open.

Traditionalists leveled a variety of charges against theorists who, they felt, had overestimated the age of the human race. The Traditionalists reiterated, for example, Forbes and Phillips's charges that Lyell's work rested on specious assumptions. It must be acknowledged, said an anonymous writer in the *British Quarterly Review*, "that the doctrine of 'uniformity' of action must be accepted with great limitation. The proof of this may be found in every work on geology, where effects are described which no amount of time can account for, without adding the other elements of *force* in varying degree."[36] Does Lyell not know, the *British and Foreign Evangelical Review* asked, "that the very best geologists deny his principle of the uniformity of geological movements?"[37] The Anglican divine Henry H. Wood offered a more succinct statement, deriding what he called "an unreasoning apotheosis of the uniformity of nature."[38]

Lyell bore the brunt of this type of criticism, but Traditionalists applied it to all who assigned humans an absolute age higher than 8,000 years. They took the fact that "geologists of the highest eminence are divided as to whether the data indicate periods of two, three, twenty, thirty, or one hundred thousand years ago" as evidence that the age of the Post-Pliocene strata could not be calculated accurately. Any attempt to do so was to use an "arbitrary hypothesis" to interpret "imperfect data."[39]

Traditionalist commentators also revived doubts about the validity of the data on which the new case for human antiquity rested. They suggested that some so-called stone tools might be

36. "Modern Anthropology," *British Quarterly Review* 38 (1861): 482.
37. "The Antiquity of Man," *British and Foreign Evangelical Review* 18 (1869): 296.
38. [H. H. Wood], "Lake Dwellings of Switzerland," *Contemporary Review* 4 (1867): 392.
39. "Modern Anthropology," 478, 483. S. R. Pattison's *New Facts and Old Records: A Plea for Genesis* (1868?), took a similar position.

naturally fractured rocks and that others might be forgeries. They questioned whether the bones and human artifacts found together at Abbeville, Brixham Cave, and similar sites might have been brought together accidentally. They asked why fossilized human bones were so scarce in areas that had yielded large numbers of stone tools. A number of commentators had raised the same questions in 1859–60, but after 1863 the Traditionalists put the questions to different use.

When first presented, in 1859–60, the arguments were meant as grounds for outright dismissal of the new case for human antiquity. Only a few, ultraconservative Traditionalists sought to use them for that purpose after 1863.[40] The majority of Traditionalists used the arguments to raise doubts about the quality of the data on which all arguments for an absolute age greater than 6,000 years rested. The result was that, in the words of one Traditionalist, "there is an indistinctness and uncertainty about these memorials of a primitive race precisely in the very particulars where distinctness and certainty are most to be desired." The author was quick to note that some of this indistinctness was due to the ravages of time. Nevertheless, he found it remarkable that "all the remains of the pre-Adamite host should be so obscure in form, and so conjectural in purpose, that a man might, without the slightest stigma upon his sagacity, use them as materials for Macadamizing a turnpike road."[41]

Warnings that scientists theorized incautiously—that there was no conclusive evidence for humans older than 6,000 years— were the core of most Traditionalist works. A few Traditionalist authors went further, however. Appealing to their readers' common sense, they argued that the "uncurbed guesses" and "vague . . . wild conjectures" of established scientists neglected obvious, contradictory facts.[42] One 1867 article noted that humans had, within a few centuries, made several species extinct and severely reduced the ranges of others.[43] Several other writers pointed out

40. See, e.g., S. Lysons, *Our British Ancestors* (1865), i–xvi; and W. Cooke, *Fallacies of the Alleged Antiquity of Man Proved* (1872).

41. "The Antiquity of Man," *British Quarterly Review* 37 (1863): 438.

42. "Modern Anthropology," 478.

43. W. C. "The Antiquity of Man," *British and Foreign Evangelical Review* 16 (1867): 383–400.

that the frozen carcasses of mammoths had been found, more or less intact, in Siberia. The excellent preservation of the mammoth corpses rendered it absurd, Traditionalists argued, that the beasts had lived in the distant past. The ferocity of human hunters explained their extinction without recourse to the gradual climatic changes demanded by geologists. Mammoths must, therefore, have lived quite recently—within the 6,000 years traditionally regarded as the human era.

THE PREADAMITE THEORISTS. A few participants in the debate over absolute age do not fit comfortably into any of the three categories above. They argued that humankind was 6,000 years old but also insisted that the age of mammoth bones and stone tools from the Drift should be calculated in tens—perhaps hundreds—of thousands of years. These commentators argued that the tools found at Abbeville, Brixham Cave, and similar sites were the work of "preadamite" races. Different authors offered different explanations of what the preadamites had been like, but all agreed on the central issue: they were earlier, and lesser, beings than the offspring of Adam and Eve.

The idea that there had been men before Adam was far from new. It had been proposed—and denounced as heresy—as early as the late sixteenth century. Nineteenth-century reactions to the idea were more restrained; most theologians and biblical scholars tolerated it, and some embraced it as an elegant solution to a variety of social, scriptural, and scientific problems. Preadamite theories, like interpretations of the six "days" of the Creation as six ages of indefinite length, depended on a growing willingness to accept a less-than-literal interpretation of the first chapters of Genesis.

Preadamite theorists noted that Genesis contains two accounts of the creation of man. In the first (Gen. 1:26–30), God creates man and woman simultaneously and gives them dominion over the Earth and all living things; in the second (Gen. 2:7–8, 18–23), God creates Adam, brings forth animals to serve and be named by him, and eventually creates Eve. Preadamites, their defenders argued, descended from the unnamed man and woman created in Genesis 1. They followed God's command to "be fruitful, and multiply" and inhabited much of the known world by the time of Adam's creation. The existence of preadamite races accounted for Cain's nameless wife, and the inhabitants of the city that he built in Nod.

Opinions varied as to how the preadamites differed from the Adamites. Some believed that the preadamites were fully human in appearance, but lacked an immortal soul.[44] A few suggested that they might have belonged to a different species than the Adamites.[45] The idea that preadamites were the ancestors of today's "inferior" (non-Caucasian) races was also popular, particularly in America.[46] The most common claim about the preadamites was simply that they were primitive. Theorists inferred, from patchy references in Genesis, that they neither farmed, herded, nor built cities, and that they were ignorant of metalworking. Those achievements—the foundations of civilization—were the work of the Adamites, the "sons of God" whose story is told in Genesis.

Preadamite theorists saw the geological and archaeological discoveries of the mid-nineteenth century as ideal corroborative evidence.[47] The ancient hunters revealed at sites throughout Europe—primitive, nomadic, ignorant of any material but stone—fit the established image of preadamite man perfectly. The Engis skull and the bones of Neanderthal Man reinforced the connection. Both were unquestionably human, but the latter showed traces of the brutish features that might mark a rough, possibly soulless being. J. Scott Moore wrote, with evident satisfaction, that "the Mosaic revelation, geological facts, and astronomical calculations (so far back as they have yet been carried) appear to agree with, and adapt themselves to, each other."[48]

Only a few participants in the human antiquity debate de-

44. J. S. Moore, *Pre-Glacial Man and Geological Chronology* (1868), 84–87.

45. J. W. Dawson, "Appendix L: Supposed Tertiary Races of Men," in *Archaia*, 2d ed. (1860), 397. There is no trace of the idea in Dawson's later work.

46. R. Popkin, "Pre-Adamism in Nineteenth Century American Thought," *Philosophia* 8 (1978–79): 205–39. D. N. Livingstone, "The Preadamite Theory and the Marriage of Science and Religion," *Transactions of the American Philosophical Society* 82, no. 3 (1992).

47. D. M'Causland, *Adam and the Adamite* 3d ed. (1872), is a particularly detailed attempt to reconcile the pre-Adamite theory with contemporary science.

48. Moore, *Pre-Glacial Man*, 40–41.

fended the preadamite theory in detail. Many clergymen and a few devout scientists alluded to it, however, noting that it offered a way of reconciling the new case for human antiquity with orthodox readings of Genesis.[49] Ascribing the stone tools from the Drift to preadamites divorced the issue of the tools' absolute age from religious issues. The tools and their primitive makers could be as old as science cared to make them. Since they predated Adam, their age was irrelevant to his. The preadamite theory thus defused science's latest challenge to the accuracy of the Old Testament genealogies. At a time when the truth of the Bible itself seemed to be at stake, such theories were matters of high importance.

The Impact of Human Antiquity

The establishment of human antiquity challenged long-held ideas about the past. The extent to which it changed those ideas and forced educated Victorians to come to a new understanding of the past is more difficult to define. The new case for human antiquity proposed by Prestwich and his colleagues touched four central areas of Victorian thought: specific theories about the origin of the human species and its constituent races; general ideas about "man's place in nature"; ideas about the relationship between science and scripture; and definitions of the boundaries and subject matter of sciences that studied the past. It had its most profound and lasting impact in the fourth of these areas. Its impact on the other three was buffered by a variety of factors, some external, others internal to the human antiquity debate. Chapter 7 analyzes how the establishment of human antiquity redefined both geology and archaeology; in the balance of this chapter I examine its impact on other aspects of Victorian thought.

Scientists discussing the problem of human origins paid close

49. See, Young, *Modern Scepticism,* 175–85; Dawson, "Appendix L," 397; "Antiquity of Man," *British Quarterly Review,* 439–41; "The Science of Anthropology," *Eclectic Review* 15 (1868): 471; "Review of *Man's Age in the World*," *Journal of Sacred Literature* 8 (1866): 478; C. J. D'Oyly, "Man in Creation," *Contemporary Review* 8 (1868): 550–68. Preadamite theories appealed to some writers on scriptural as well as scientific grounds. See *The Genesis of the Earth and of Man,* ed. R. S. Poole (1860), and the anonymous review of it in the *Journal of Sacred Literature* 12 (1860): 123–33.

attention to the new case for human antiquity. The new case contributed to the long-standing debate over monogenism and polygenism and to the emerging debate over human evolution, but only in subsidiary roles. It was more often a means of reinforcing existing theories than a spur to the formation of new ones. The lack of a consensus on absolute age meant that scientists could take virtually any position that suited them and still claim the support of recognized geological authorities. Free from constraints imposed by nature or their colleagues' consensus, scientists adopted views on the absolute age of the human race that paralleled and reinforced, rather than challenged, theories about humankind that they already held.

Consider, for example, the impact of human antiquity on ideas about human evolution. Huxley, Wallace, and other evolutionists readily accepted the idea that humans had lived in the Post-Pliocene. They adopted, then quickly expanded, Lyell's 100,000-year figure for the human race's absolute age. This ready acceptance of an estimate that many commentators deemed excessive or even absurd reflected both the evolutionists' acceptance of hypotheses as useful tools and evolutionary theory's demand for vast spans of time. It is clear, however, that the evolutionists regarded geological evidence of human antiquity simply as partial confirmation of what biology had already demonstrated. The flints from Brixham Cave and Abbeville buttressed, but did not expand, the fledgling case for human evolution.

The same pattern is apparent in other parts of mid-Victorian science. Monogenists adopted Lyell's expansive views on absolute age because the development of the races from a single stock demanded time.[50] Polygenists, such as John Crawfurd and James Hunt, took the same position for the opposite reason. They offered the presence of the "fully modern" Engis skull in 100,000-year-old deposits as proof of their belief that the human form did not, and could not, change.[51] Opponents of human evolution of-

50. [R. G. Latham], "Antiquity of the Human Race," *Westminster Review* 74, American ed. (1860): 219–33.

51. J. Crawfurd, "On Lyell's 'Antiquity of Man,'" *Transactions of the Ethnological Society of London*, n.s., 3 (1865): 58–70; J. Hunt, "The Principles of Archaic Anthropology," *Popular Magazine of Anthropology* 3, appended to *Anthropology Review* 4 (1866): 89–93.

ten favored a Prestwichian or Traditionalist position, probably because Darwin's theories could not account for the origins of a
species only twenty thousand years old. The dwindling scientific
rear guard who sought to reconcile geology with Genesis took
similar positions. Scientists' answers to the absolute age question
also reflected their views on method. Older men such as John
Henslow, committed to a strict Baconian approach, tended to
adopt Prestwich's view of the Somme Valley evidence and to reject Lyell's view as too far removed from the data.

Most educated Victorians who commented publicly on the
human antiquity problem were familiar with these scientific issues. They were less concerned with particular theories, however, than with the theories' broader implications. Victorians' understanding of "man's place in nature"—and consequently of
God's place in nature—changed profoundly in the fifteen years
after 1860. Developments in science and religion challenged and
ultimately transformed it. The 1859 publication of Darwin's *Origin of Species* was the most visible challenge, but it was aided and
abetted by a wide range of less obvious ones. Victorians outside
the scientific community approached the new case for human
antiquity with those challenges clearly in mind.

The satirical poem "Monkeyana," which appeared in the May
18, 1861, issue of *Punch*, indicates the human antiquity question's prominence in the minds of educated Victorians. Signed
"GORILLA, Zoological Gardens," the poem was illustrated by an
engraving of a very human-looking ape standing upright and
wearing a sign that asked "Am I A Man and A Brother?" The
simian "author" posed a similar question in the first verse:

> Am I satyr or man?
> Pray tell me who can,
> And settle my place in the scale.
> A man in ape's shape,
> An anthropoid ape,
> Or monkey deprived of his tail?[52]

52. Gorilla, "Monkeyana." The full text of the poem appears immediately
preceding chap. 1. "Monkeyana" was the work of Sir Philip Egerton, one of Britain's most respected paleontologists and former president of the Geological Society. Huxley attributes the poem to him in a letter to Joseph Hooker, printed in L.
Huxley, *Life and Letters of T. H. Huxley* (1901), 1:206–7. The poem also reflects ideas

Succeeding verses touched on the *Vestiges of Creation*, Darwin's theory of natural selection, and T. H. Huxley's dispute with Richard Owen over the anatomical similarities of humans and apes. Three verses were devoted to the human antiquity problem, including one on Pengelly's excavation of Brixham Cave and another on Prestwich's investigation of the Somme Valley gravels.

A single poem cannot adequately capture the mind-set of an entire nation, but "Monkeyana" was more than an amusing piece of ephemera. The oblique references to Prestwich's and Pengelly's work suggests that the average reader of *Punch* was already familiar with the evidence for human antiquity. The extensive use of technical terms in the Huxley-Owen verses supposes that the reader was also familiar with the evidence for human evolution. The casual juxtaposition of biological, geological, and anatomical issues implies that the connections between them were already common knowledge among educated Britons. Finally, the appearance of such a poem in a magazine little concerned with scientific progress suggests that man's place in nature had become a topic of general discussion by 1861, and that human antiquity was a significant part of that discussion.

John Lyon has shown that, in the 1820s and 1830s, opposition to theories of human antiquity stemmed partly from opposition to the idea that humans might have lived on an "imperfect" Earth. Scientists and laymen alike interpreted the fossil record teleologically. They saw the increasing complexity of flora and fauna as evidence that God had gradually prepared the Earth to receive His finest creation: man. The extinction of Europe's mammoths, cave bears, and wooly rhinoceroses thus marked the end not only of the Post-Pliocene epoch but also of this preparation process. The presence of humans among these now-extinct animals implied that humans had been placed on a still-incomplete Earth, and threatened a view of the fossil record that had been popular among clergymen, lay people, and devout scientists alike.[53]

---

and turns of phrase that Egerton used in a letter written to Louis Agassiz the previous month (Egerton to Agassiz, April 16, 1861, in E. C. Agassiz, *Louis Agassiz: His Life and Correspondence* (1885), 2:571–72).

53. J. Lyon, "The Search for Fossil Man," *Isis* 61 (1970): 68–84.

Donald Grayson notes that the discoveries made in 1858–59 overwhelmed objections to humankind's presence on an "incomplete" Earth, and that the evidence marshaled in Lyell's *Antiquity of Man* extinguished them entirely.[54] Acceptance of the case for men among the mammoths did little, however, to upset the idea that man was the crowning glory of God's creation. William Whewell, though regretful about the blurring of the human era's once-clear boundaries, remained optimistic. Reconciliation of scientific and religious ideas about the past was still possible, he wrote to geologist James Forbes, though "it is not so clear and striking as it once was."[55] Whewell concluded that the basis for such a reconciliation would soon be found; in fact, it was already under way. A number of writers seamlessly integrated the evidence for human antiquity into traditional arguments that the Earth had been prepared for its human inhabitants. The more imaginative among them used human antiquity to expand on the idea.

Books and articles on human antiquity appearing after 1860 regularly emphasized the continuity of late Post-Pliocene Europe and modern Europe. Their authors argued that the geological changes separating the Post-Pliocene from the Recent epoch had been relatively minor. The details of this argument varied from author to author, but Joseph Prestwich presented a well-detailed, oft-quoted version. The Abbeville toolmakers, Prestwich wrote in 1859, had known a landscape similar in all but minor features to that of the present day. They had lived amid vegetation "very similar to that now flourishing in like latitudes," and among mammals "of which some species are yet the associates of man." He emphasized—and, in an 1862 paper, reemphasized—that "an uninterrupted succession of life" connected the Post-Pliocene epoch with the modern one."[56] The Earth, in other words, had been nearly as "finished" in the days of the Abbeville toolmakers as it was today.

54. D. K. Grayson, *The Establishment of Human Antiquity* (1983), 201–5.

55. W. Whewell to James D. Forbes, January 4, 1864, in I. Todhunter, *William Whewell, D. D.* (1876), 2:435.

56. J. Prestwich, "On the Occurrence of Flint Implements," *Philosophical Transactions of the Royal Society* 150 (1860): 308–9; Prestwich, "Theoretical Considerations," 303–5.

Commentators who differed sharply on the absolute age question concurred that humans had been the last addition to a nearly complete Earth. Dominick M'Causland argued that the age of the human race is "long posterior to the age of the now existing flora and fauna, or even to the time of their actual geographic distribution. Man, whatever may have been the date of his first appearance upon the earth, was undoubtedly later on the scene than any other item of the creation. All classes and orders of organisms below the rank of humanity were in existence when man was not."[57] The pseudonymous "Essex Rector" concurred that humans "did not appear until the face of Europe was much the same as it is now." The *British and Foreign Evangelical Review* noted that, though human history might have been expanded, it still belonged to the very last stage of geologic time. The advent of humankind remained "an event, comparatively speaking, of yesterday."[58]

Commentators who took this approach did not dispute the idea that the appearance of humankind marked a turning point in Earth's history. They simply shifted the turning point backward in time. The end of the glacial period, rather than the extinction of the mammoth and similar beasts, became the favored geological marker for the beginning of the human period. Geologists of the 1860s and 1870s equated the glacial period with roughly the first half of the Post-Pliocene epoch. That, and the fact that every human bone and artifact then known had come from postglacial deposits, made the end of the glacial period a convenient time marker.

The glacial period also offered other, more abstract advan-

57. M'Causland, *Adam and the Adamite*, 47–48.

58. Essex Rector, *Man's Age in the World*, 15; "Recent Geological Speculations," *British and Foreign Evangelical Review* 10 (1861): 885. Lyell's journal entries on the human antiquity problem show that he, too, regarded humankind as part of the "yesterday" of geological time. See L. G. Wilson, *Charles Lyell's Scientific Journals on the Species Question* (1970), 242, 246. For similar published views see duke of Argyll, "Opening Address," *Proceedings of the Royal Society of Edinburgh* 4 (1862): 363; Dawson, "Primitive Man and Revelation," 59–61; [W. F. Pollock], "Review of *Antiquity of Man*," *Fraser's* 67 (1863): 472; D. T. Ansted, "The Testimony of Geology to the Age of the Human Race," in *Geological Gossip* (1860), 304–11. The authorship of "Testimony of Geology" is inferred from Ansted's comments in the preface of *Geological Gossip*.

tages. It had, according to most geologists, accounted for the last significant changes in the geology and topography of Western Europe—changes that could be seen as final preparations for the appearance of the first humans. Dawson envisioned the end of the glacial period as "giving new features of beauty and utility to the land, and preparing the way for the life of the Modern period, as if to make up for the time which had been lost in the dreary Glacial age." Prestwich, in an 1862 paper, argued that the glaciers had cooled and solidified the Earth's crust, arresting its cyclical up-and-down motion and rendering it "fit and suitable for the habitation of civilized man." This, he felt sure, was evidence of "great and all-wise design."[59]

Some commentators preferred to use biological events to mark the appearance of humans on Earth. John Phillips's 1860 paleontology textbook *Life on Earth* noted that humankind and the animals most beneficial to it appeared on Earth simultaneously. It is clear, given Phillips's earlier statement in the same book that "Nature, in a large sense, is the expression of a DIVINE IDEA," that he saw the simultaneity as evidence of providential design. Samuel Haughton, professor of geology at the University of Dublin, was one of several authors who argued that humans— placed on an all-but-modern Earth—had made it fully modern by hunting its large, once-dominant mammals to extinction.[60]

These arguments supported a single, reassuring conclusion: Humans were something special in the history of life. They had been the last species to appear on Earth, and they had done so in the very last stages of the history of life. Their appearance had marked the beginning of a distinct, recognizably modern geological epoch and had been preceded by events apparently designed to better suit the Earth to their needs. This, combined with the idea that ancient hunters had killed off the mammoth and

59. Dawson, *Story of the Earth and Man,* 287; Prestwich, "Theoretical Considerations," 305.

60. J. Phillips, *Life on Earth* (1860), 50, 3; S. Haughton, "History of the Earth and Its Inhabitants," *Dublin University Magazine* 58 (1861): 112. For other views along the same lines, see: Owen, *Paleontology,* 401; M'Causland, *Adam and the Adamite,* 45–47; and W. C., "Antiquity of Man." D. K. Grayson ("Nineteenth-Century Explanations of Pleistocene Extinctions," in *Quaternary Extinctions,* ed. P. Martin and R. Klein [1984]) discusses the popularity of extinction-by-early-man theories.

other once-dominant species, reinforced traditional beliefs in man's God-given dominion over nature. The adjustment in humankind's relative age forced by the new case for human antiquity was only minimally troubling when seen in such a light. There were, in any event, greater threats to face.

The expansion of scientific knowledge in the first half of the nineteenth century seriously depleted the stock of phenomena that could not be attributed to natural causes. British scientists and clergymen who wished to demonstrate the presence of God in nature increasingly sought evidence of His provident design, rather than of His direct intervention. The special creation of new species—each well-adapted to its environment—was, by the 1850s, perceived as the last venue for God's continued, intimate involvement in the natural world. Darwin's *Origin of Species*, published in November 1859, evicted God from the speciation process in favor of natural, nonteleological forces. Thus, it threatened not only to eliminate the last strand of evidence for God's direct intervention in nature but also to demolish the argument from design. Nor did it stop there. Though Darwin touched on the issue only briefly, his book raised the horrifying possibility that humans, too, were simply a product of natural forces.[61]

The public outcry provoked by *Origin of Species* was, as Ellegard has documented, stimulated largely by the prospect of an evolutionary origin for man. Outside the scientific community, Darwinian evolution was widely referred to as "the monkey theory." The humanlike gorilla that headed "Monkeyana" was only one of many "ancestors" whose portraits appeared in *Punch* and similar periodicals.[62] Commentators willing to accept the possibility that the lower animals had been created by natural causes balked at extending the argument to humans.[63] Commentators who discussed both human antiquity and human evolution invariably regarded the latter as a more serious threat to orthodox Christianity. An anonymous author in the *British Quarterly Review* wrote, "If we were, by force of inevitable logic, compelled to ex-

---

61. C. C. Gillispie, *Genesis and Geology* (1951), chap. 8; Moore, *Post-Darwinian Controversies;* Ellegard, *Darwin and the General Reader.*

62. Ellegard, *Darwin and the General Reader,* 154–65, 293–99 ff.

63. Bartholomew ("Lyell and Evolution") traces one excellent example—Lyell's personal struggle with the idea of human evolution.

tend our ideas of the length of our existence as a race by some thousands of years, we might do so . . . without necessarily endangering the entire structure of our faith, as founded upon a revelation. This cannot be said concerning the next question that must engage our attention: viz., that of the Origin of man."[64]

Anglicanism was challenged from within, as well as without, during the 1860s. Even as Victorians struggled to come to terms with *Origin of Species,* the multiauthored volume *Essays and Reviews* appeared in 1860. *Essays and Reviews* was the manifesto of the Broad Church movement—an informal alliance of Anglicans who believed that both biblical texts and the authority of the church could be subjected to systematic criticism. The book's authors, particularly the Reverend Benjamin Jowett, brought the Anglican establishment and the public face to face with a style of scriptural analysis developed in Germany decades earlier. "Interpret the Scripture like any other book," Jowett declared to his readers; explicate it not as a sacred text, but as a secular one. Elsewhere in *Essays and Reviews,* the Reverend Baden Powell applied Jowett's principles to the miracles described in the Old Testament. Modern science, he concluded, could readily explain what the unsophisticated Israelites three millennia before could attribute only to acts of God.[65]

The furor generated by *Essays and Reviews* intensified as the decade progressed. J. W. Colenso, the Anglican bishop of Natal, South Africa, published his own contribution to German-style "higher criticism" in 1862. Colenso's book, *The Pentateuch and the Book of Joshua Critically Examined,* followed Powell and Jowett in reading the Old Testament as the work of fallible human authors whose work was the product of a specific place and time. It suggested that familiar biblical stories such as those of Noah and Jonah might be fables, metaphors, or products of poetic exaggeration. Like *Essays and Reviews,* it left many readers convinced that

---

64. "Modern Anthropology," 483.
65. O. Chadwick, *The Victorian Church* (1967–70), 2:74–90. For a more detailed discussion, see M. A. Crowther, *Church Embattled* (1970). For the impact of *Essays and Reviews* on the British scientific community, see F. Gregory ("The Impact of Darwinian Evolution on Protestant Theology in the Late Nineteenth Century") and J. R. Moore ("Geologists and Interpreters of Genesis in the Nineteenth Century"), both in *God and Nature,* ed. D. Lindberg and R. Numbers (1986).

the inspired nature of the Bible—and so the basis of Christianity—was under attack.[66] Colenso, stripped of his bishopric in Natal, fled to London, where he found himself welcomed by Lyell, Huxley, and other leading scientists.

Essays and Reviews and Colenso's *Pentateuch* masked the impact of human antiquity on biblical interpretation, just as *Origin of Species* had masked its impact on ideas about man's place in nature. Scientists' claims that humankind might be more than 6,000 years old implied, at worst, that parts of Genesis might need to be reinterpreted. The work of Colenso, Jowett, and Baden Powell called *all* traditional interpretations of the Bible, as well as the principles on which they rested, into question. Many regarded the higher criticism as an attack on the divinely inspired status of scripture, and so on the basis of Christian morality.

The threat to Christianity from the higher criticism was new and strange and struck at the foundations of orthodox belief. Conflicts between the details of Genesis and of geology were, by the 1860s, neither new nor strange. They had arisen before and, in the eyes of most, had been resolved. The growth of geological knowledge had shown that the Earth was not 6,000 years old, that life had not appeared in a single week, and that there had been no worldwide flood in roughly 2,000 B.C. The challenge posed by the new case for human antiquity was, therefore, the latest chapter in a familiar story. It was met with time-tested strategies that were familiar to scientists, clergy, and laymen alike.

The new case for human antiquity posed, in some respects, a comparatively minor challenge to traditional readings of Genesis. It was not clear, in the absence of a scientific consensus on the human race's absolute age, that a conflict even existed. Commentators often argued that until science could rest its absolute age calculations on a solid foundation of evidence, any reinterpretation of scripture would be premature. J. W. Dawson's well-publicized belief that the stone tools from the Drift might be only 6,000 years old gave scientific weight to this position. The calls for cautious theorizing published by H. D. Rogers and the Duke of Argyll also reinforced it, though perhaps not by design.

66. Chadwick, *Victorian Church*, 2:90–97.

Many of those who did see a conflict turned to preadamite theories as an effective method of harmonizing old readings of Genesis with the new evidence for human antiquity. Harmonization, an intellectual endeavor well established by the mid-nineteenth century, rested on the assumption that the language used in Genesis was poetic rather than precise, and that clarifying its meaning did not distort its message. Like earlier harmonizing schemes that saw the six "days" of creation as geological ages and Noah's flood as a regional inundation, preadamite theories preserved the central theme of the Creation story at the expense of some details. They required a highly imaginative reading of certain verses but left unchanged all direct references to Adam and his offspring. Adam, the preadamite theorists were careful to note, was the first true man: God created him, in His image, to rule the Earth. The descendants of Adam—not those of the savage preadamites—farmed, herded, built cities, worked in metal, and received the Ten Commandments.

Preadamite theories, like claims that the new case for human antiquity required no reinterpretation of Genesis, were most popular among amateur interpreters of Scripture. Scientific support for them came primarily from a few aging career scientists and a large body of amateurs. The younger, newly professionalized members of both the scientific community and the clergy rejected, on philosophical grounds, the idea that Genesis and geology needed to be harmonized. Science and religion, they argued, constituted separate intellectual spheres. One was concerned with the natural world, the other with the moral and spiritual world; attempts to reconcile one with the other distorted both and did justice to neither.[67] Malcolm White's 1869 defense of this position was titled "Does Scripture Settle the Antiquity of Man?" Its message was clear: the answer was, "No." White dismissed the need for "feverish anxiety about the result of the next [scientific] discoveries" and for "shallow hypotheses respecting the mythical character of the early chapters of the Bible." Those chapters, he concluded, "are concerned with man's fall and the means of his salvation; but as to curious questions about when he fell and [when] the race was scattered abroad, they seem to maintain si-

67. Moore, "Geologists and Interpreters of Genesis," 328–45.

lence, as the rest of Scripture does about many other topics on which science spends its utmost labor."[68] Professional biblical scholars ignored, as White did, the exegetical problems that many amateurs believed the new case for human antiquity had raised. They believed, as professional scientists did, that the problems were illusory.

TWO THINGS masked the impact of human antiquity on Victorian thought: the timing of the major discoveries and the structure of the case that geologists built around them. The papers announcing the geologists' new evidence and arguments for human antiquity appeared in spring 1859, only months before *Origin of Species* and less than a year before *Essays and Reviews*. Lyell's *Antiquity of Man*, which brought the new case before the educated public, appeared only months after Colenso's *Pentateuch* and almost simultaneously with Huxley's *Man's Place in Nature*. The public debate over human antiquity, which reached its peak in the mid-1860s, was overshadowed by the controversies surrounding evolutionary theory and the "higher criticism" of the Bible. Claims about the age of the human race—that might in other times have seemed shocking—were tame compared to theories that seemed to derive men from monkeys and the Bible from fallible human authors. Defenders of orthodoxy, besieged from all sides, fought what they saw as the greater evil and made a separate peace with the lesser.

The structure of the new case for human antiquity made such a separate peace possible. The presence of men among the mammoths raised questions about man's place in nature but none that could not be answered within the context of existing ideas about the past. The lack of a scientific consensus on the absolute age issue also made the evidence for human antiquity easier to accept. Science offered a wide range of answers to the question "How many years ago did the first Europeans live?" allowing scientists and laymen alike to choose the one that suited them best.

The geologists' case for human antiquity was flexible enough

68. M. White, "Does Scripture Settle the Antiquity of Man," *British and Foreign Evangelical Review* 21 (1869): 136–37.

to accommodate traditional views of the past—whether scientific or religious—with only minor modifications. It changed the details, but not the core assumptions, of Victorian ideas about man's place in nature. The new case's impact on the boundaries and content of scientific disciplines that studied the past was another matter. The boundaries that had kept geology, archaeology, and anthropology distinct rested on the assumption that humans belonged exclusively to the Recent epoch. The establishment of human antiquity forced a redrawing of disciplinary boundaries by demolishing the assumption on which the old ones rested and raising questions that Victorian science had never asked before. The result was a revolution in British science and a new, multidisciplinary approach to the study of the geologically recent past.

SEVEN

New Approaches to Prehistory

THE BARRIER that had once separated the human era from the era of extinct animals began to crumble in 1858–59. The last vestiges of its disappeared by 1863, and with them the intellectual foundation on which traditional definitions of geology and archaeology rested. English geology may have become a historical science through the interaction of seventeenth-century naturalists and antiquaries,[1] but by the eighteenth century the two disciplines had become clearly and distinctly separated. Archaeology began, at least in principle, where geology ended. Geologists had carefully excluded humans from their studies, and archaeologists had focused their attention on the few thousand years illuminated by document and tradition. The establishment of human antiquity showed that the subject matter of geology and archaeology were interleaved. It revealed unsuspected chapters in the early history of the human race—chapters that archaeology could not study in isolation from geology.

Prehistory existed in British archaeology, as both a concept and a word, well before the excavation of Brixham Cave began. The new case for human antiquity did not reveal its existence for the first time. The new case showed, however, that prehistory was both larger and more complex than British archaeologists had ever suspected. In the 1850s, they had measured the prehistory of Europe in centuries; from the 1860s onward, they would measure it in tens—sometimes hundreds—of thousands of years. It was, in Thomas Trautman's evocative phrase, as if "the bottom dropped out of history." The new case for human antiquity also

1. C. J. Schneer, "The Rise of Historical Geology in the Seventeenth Century," *Isis* 45 (1954): 256–68.

established, for the first time, that the Celts had not been the first inhabitants of Europe. It revealed a race of primitive hunters that had spread over southern and western Europe as the glaciers retreated, leaving their primitive tools behind among the bones of their prey.

Together, these discoveries transformed British archaeologists' ideas about prehistory. What had been a comparatively brief prelude to the history of the later Celts, the Romans, and the Saxons became many times longer than all of written history. The stone tools found in the Drift reflected nothing so clearly as their makers' primitiveness. The archaeological implications of human antiquity were clear: the earliest inhabitants of Europe had lived in abject savagery for thousands upon thousands of years.

British archaeologists possessed, in 1859, neither the inclination nor the methods to study prehistory in detail. The challenge posed by the establishment of human antiquity caused both an internal and an external restructuring. Archaeology had, in Britain, been a socially and intellectually unified discipline guided by a research tradition I have labeled "historical archaeology." It split, by the mid-1860s, into two subdisciplines that had little social or intellectual overlap. The majority of established archaeologists remained a cohesive body and remained committed to historical archaeology's methods, research agenda, and geographically insular outlook. The remainder, joined by scientists new to archaeology, formed a second, entirely new subdiscipline and established a research tradition of their own that borrowed methods and goals from the natural sciences.

The founders of the new subdiscipline referred both to it and to the research tradition associated with it as "prehistoric archaeology." The two were, in their minds, synonymous. A century and a quarter later, however, it is important to distinguish between them. Though prehistoric archaeology is still a flourishing specialty within the larger science of archaeology, the methods and assumptions current in the 1860s have long since been supplanted.[2] The research program founded by Lubbock and his col-

2. The precise relationship between prehistoric archaeology's past and present research programs is an open question. See C. Chippindale, "'Social Archaeol-

leagues drew its methods and assumptions from several of the natural sciences, but most conspicuously from geology. I will therefore, for the sake of clarity, refer to it as *geological archaeology* and to its founders as *geological archaeologists.*

Geological archaeologists became the geologists' new partners in the study of the earliest humans. Geologists and geological archaeologists pursued very different research agendas, but they regularly shared data and frequently cooperated in excavations and other fieldwork. The geological archaeologists' writings reflected their familiarity with the facts of geology, as well as their use of its methods. Geological archaeology also became closely allied with the emerging science of anthropology. Lubbock and Pitt-Rivers worked in both fields, drawing analogies between their studies of ancient artifacts and of "modern savages."

The founders of the new subdiscipline and its research tradition—John Evans, John Lubbock, Augustus Pitt-Rivers, and William Boyd Dawkins—were all young men.[3] They came from disparate intellectual backgrounds but shared an interest in prehistory and a belief that it could best be studied by using techniques borrowed from geology. Their work drew extensively on geological and anthropological data and often appeared in the journals of geological and anthropological societies. They saw themselves as archaeologists, but they had much closer social and institutional ties with geologists and anthropologists than with the historical archaeologists.

Geological archaeology's intimate ties to anthropology and geology, along with its practitioners' belief in international cooperation, quickly established it as an integral part of Victorian science. Its contributions to emerging debates over human biologi-

---

ogy' in the Nineteenth Century," in *Tracing Archaeology's Past,* ed. A. Christenson (1989).

   3. Pitt-Rivers was born Augustus Henry Lane-Fox, but took a new surname in 1880 after he inherited the estates of his uncle George Pitt, the second baron Rivers. His adopted name is far better known today than his birth name, and I have used it here (despite the anachronism) for that reason. John Lubbock was elevated to the peerage as Lord Avebury in 1901. Writings by him and about him that appeared between 1900 and ca. 1970 tend to refer to him as "Avebury," while later writings tend to use "Lubbock."

cal and cultural evolution gave it notoriety. These men became prominent in the British scientific community both as officers in leading societies and members of informal but influential networks. The international prominence of geological archaeology, combined with the gradual marginalization of the historical archaeologists by historians, altered the balance of power in the archaeological community. Historical archaeology remained an immensely popular pursuit, but geological archaeologists' studies of prehistory defined the cutting edge of British archaeological research.

Historical Archaeologists and Prehistory, 1860–80

Historical archaeology claimed all of British history, whether recorded in documents or only in legends, as its subject matter. The historical archaeologists' intellectual domain reached, in principle, from the Celts to the late seventeenth century. In practice, however, some parts of that domain were destined to be far better explored than others. Historical archaeology was at its best when analyzing the details of life in medieval or Renaissance Britain; the archaeologist could gather a wealth of facts—from artwork, legends, archaic words, and material artifacts—and fit them into a chronological framework based on written documents. Given sufficient time and effort, he could create a detailed picture of a vanished age without ever straying from "true Baconian principles." The picture, when finished, would illuminate a short but recognizable stage in the development of the British people.

The earlier Roman and Saxon periods offered the archaeologist fewer and less varied artifacts, virtually no written documents, and little hope of producing a complete, detailed reconstruction. The shadowy Celtic period offered no hope at all; the Celts had left behind only stone and bronze tools, some jewelry, and hundreds of mysterious monuments. Not surprisingly, historical archaeologists focused their efforts on the Middle Ages. The contents of their national journals reflected a modest but persistent effort to understand Roman Britain, a tenuous interest in the Saxon period, and little more than an attempt to catalog the remains of the Celts. The images historical archaeologists used to describe the Celtic period—mist, fog, and shadows—suggest a time not only unknown but largely unknowable.[4]

The archaeological community's labeling of the Celtic period as chronological terra incognita passed unchallenged during the 1840s and 1850s. The human race, according to leading geologists, was a recent addition to the Earth, and since humans had arisen in Asia, the first humans to live in Europe were more recent still. Thus, the Celtic period of British history encompassed only the time between the arrival of humans in Europe and the Roman invasion of 55 B.C.—perhaps a few centuries. Such a period was not dauntingly long when measured against the entire span of recorded British history. It could be consigned to the realms of the unknowable without raising serious theoretical or philosophical problems and, opaque to standard archaeological methods, it often was.

The establishment of human antiquity made a shambles of this tidy picture of the past. It did not extend the span of written history, but it stretched European prehistory from a few hundred to many thousands of years. The absence of a consensus on how *many* thousands of years prehistory encompassed mattered little to historical archaeologists. What did matter was that prehistory—the era of the Celts, and the era of the chipped-stone-tool makers that preceded it—now encompassed more time than the Roman, Saxon, and medieval eras combined. The majority of humankind's time on Earth now lay beyond the reach of written documents and chiseled inscriptions, represented only by mute Celtic monuments and crude stone implements such as those from Brixham Cave and Hoxne.

These changes made it impossible to dismiss prehistory as an inconsequentially small part of humankind's time on Earth. They also presented historical archaeologists with the daunting task of understanding a people even more mysterious than the shadowy, mist-shrouded Celts. The material remains of Europe's pre-Celtic inhabitants were sparse and unvaried; the Drift had yielded a few types of stone tools, but no pottery, jewelry, or domestic objects.

4. Examples include W. O. Stanley, "On the Remains of Ancient Circular Habituations in Holyhead Island," *AJ* 24 (1867): 229–64; H. S. Cuming, "On an Ancient British Snow Knife," *JBAA* 24 (1868): 125–28; and W. C. Borlase, "Account of the Exploration of Tumuli at Tevalga;" *Arch.* 24 (1873): 422–27.

Post-Pliocene toolmakers had left no buildings or monuments—not even the burial mounds, cromlechs, or simple upright stones with which Celtic tribes had dotted the landscape. Their artifacts could not be collected by systematically excavating graves or "turning over" barrows, but had to be sought in caves, river beds, and other sites more familiar to geologists than archaeologists. Even then, the Post-Pliocene toolmakers left no marks on the landscape to signal their long-ago presence. Archaeologists had no hope of reckoning how long the toolmakers had lived in Europe because they lay beyond the reach both of history and tradition. No eyewitness—friend or conqueror—had ever written down impressions of the Europe's pre-Celtic peoples.

Understanding the Post-Pliocene toolmakers presented the historical archaeologists with major methodological problems and promised few scientific rewards. "True Baconian principles" were unlikely to produce more than a list of the toolmakers' scattered artifacts. Using hypotheses to give order to and bridge the gaps between artifacts would violate those principles and still might not create a clear picture of Post-Pliocene life in Britain. Even if such a picture did emerge, what would be its value to scientists committed to understanding the roots of modern British institutions in the past? If the toolmakers of Hoxne and Brixham Cave had played a role in the creation of the British nation, it was a minor one. If there was something uniquely "British" about their culture, their stone tools were unlikely to reflect it. To the historical archaeologists, the people of the Post-Pliocene were too old to be understood or even to be particularly interesting.

Historical archaeologists did not ignore the chipped-stone tools from the Drift, but they made no concerted effort to understand the toolmakers. The national archaeological journals printed occasional descriptive papers about the tools or reports of new discoveries but little else on the subject. The leading societies convened meetings to study the new type of artifact, but these, too, involved little more than the display and description of individual tools. George Tomline declared, in his 1865 presidential address to the British Archaeological Association, that "the flint implements of Hoxne—the production of an age so distant, that, though satisfactorily demonstrated to our reason, it almost appalls our imagination—are true objects of interest to antiquaries," but historical archaeologists found the flint implements "interesting" primarily as historical curiosities.[5] An 1865 address by Cambridge's new professor of archaeology noted that archaeology began in the Quaternary period but gave fifteen pages to medieval subjects and only one to the pre-Celtic period.[6]

Historical archaeology prospered after 1860, but as it prospered it held firmly to its well-established methods and worldview. The Society of Antiquaries, the British Archaeological Association, and the Archaeological Institute all flourished, continuing to devote most of their attention to Roman, or later, remains.[7] New periodicals devoted to historical archaeology appeared: *The Reliquary* in 1860, *The Antiquarian* and *Long Ago* in the early 1870s, and *The Antiquary* in 1880. Thomas Wright brought out revised editions of *The Celt, The Roman, and the Saxon* in 1861 and 1875, and three more editions were published after his death in 1877.

The new books and journals faithfully reflected the methods and ideas of the old. Their discussions of prehistoric sites and artifacts included the Celtic period, but not the pre-Celtic. Their subject matter included a staggering range of topics; the first issue of *The Antiquary* listed fifty topics that fell "within the scope of our Magazine," including local antiquities, archaeology, bells, church furniture, heraldry, monumental brasses, provincial dialects, local traditions, and cathedrals. The same issue renewed the historical archaeologists' commitment to strict empiricism, stating that archaeology held "an honourable position among the exact inductive sciences, furnishing data subsidiary to the still grander task of the historian." The editor of *The Antiquarian* noted that only by "laborious industry in collecting facts . . . and by their comparisons," could an archaeologist form solid generalizations.[8]

Barrow digging, an activity widely practiced by the few his-

5. G. Tomline, "Inaugural Address," *JBAA* 21 (1865): 3.

6. C. Babington, *Introductory Lecture on the Study of Archaeology* (1865), 16–17, 47–62. Babington also devoted thirteen pages to classical civilizations, fifteen to Egypt and Mesopotamia, and two to the Celts.

7. See App. I for statistics.

8. G. C. Swayne, "The Value and Charm of Antiquarian Study," *Antiquary* 1 (1880): 3; "Preface," *Antiquarian* 1 (1871), quoted in K. Hudson, *A Social History of Archaeology* (1981), 100.

torical archaeologists interested in pre-Roman remains, flourished after 1860. Prehistoric archaeologists also took an interest in barrows—burial mounds that often yielded both human remains and a variety of artifacts—but historical archaeologists outnumbered and outdug them. Britain's five most prolific barrow diggers were all historical archaeologists, and three—John Mortimer, William Greenwell, and W. C. Borlase—began their multidecade careers in the 1860s. Barry Marsden's history of barrow digging lists nineteen "principal early works" on barrows, nine of which appeared after 1860.[9] Though they frequently uncovered and described prehistoric artifacts, the barrow diggers remained true to their roots in historical archaeology. They took no notice of Paleolithic artifacts which predated the barrow builders, and were at least as interested in Roman- and Saxon-era burial mounds as in prehistoric ones. As Llewellyn Jewitt insisted in 1870, their goal was to record facts, not to recreate the lives of the barrow builders or deal in other "extraneous matters."[10] The historical archaeologists who dug barrows were, in the end, no more affected by the establishment of human antiquity than were their colleagues who studied medieval churches.

Thomas Trautman has suggested that scientists' reactions to the establishment of human antiquity varied sharply with their ages. For those who began their careers in the 1840s and 1850s, he argues, it was a central intellectual event that profoundly changed their worldview. For some, like American anthropologist Lewis Henry Morgan, it was more important even than the publication of *Origin of Species*. Those who began their careers after 1860 took human antiquity for granted and incorporated it seamlessly into their own worldviews. Younger scientists, Trautman contends, found it difficult to understand just how disorienting the "time revolution" had been for their older colleagues.[11]

The structure of the British archaeological community after 1860 clearly reflected this distinction. The leading historical archaeologists, most of whom had begun their careers well before 1858, knew that the establishment of human antiquity had inval-

9. B. M. Marsden, *The Early Barrow Diggers* (1974), 116–17.
10. L. Jewitt, *Ancient Burial Mounds and Their Contents* (1870), xxiii.
11. T. R. Trautman, *Lewis Henry Morgan and the Invention of Kinship* (1987), 220.

idated many of their assumptions about the past. They knew that archaeology's subject matter had been vastly expanded but never fully assimilated the new picture of the past and altered the boundaries of their science accordingly. A significant part of humankind's early history thus lay, unstudied, in a no-man's-land between geology and historical archaeology. Beginning in the mid-1860s, prehistoric archaeology filled this newly created gap in the disciplinary landscape.

The Geological Archaeologists and Their Science

The scientists who first made prehistoric archaeology a distinct subdiscipline in Britain were all comparatively young men when human antiquity was established. John Lubbock, who became the leader of the "prehistoric movement," was 23 in 1859; William Boyd Dawkins was 22, Augustus Pitt-Rivers was 32, and John Evans was 36. All four began their archaeological careers during or soon after the initial debates over human antiquity. Evans all but gave up geology for prehistoric archaeology soon after his 1859 visit to the Somme Valley, and Lubbock's interest in prehistory emerged after a similar trip with Joseph Prestwich in 1860.[12] Boyd Dawkins, inspired by the cave explorations carried out by William Buckland in the 1820s, became a specialist in cave paleontology while at Oxford in 1859. News of the discoveries at Brixham and St. Acheul, and his own discovery of human remains and extinct animals in Wookey Hole, Somerset, soon drew him toward prehistoric archaeology.[13] Pitt-Rivers saw the extended prehistory disclosed in 1858–59 as a means of testing theories of cultural development that he had formulated as an ethnologist.[14]

The four men shared a view of the past very different from

12. Joan Evans, *Time and Chance* (1943), 82–84; A. Keith, "Anthropology," in *The Life-Work of Lord Avebury*, ed. U. Grant Duff (1924), 70–73.

13. W. B. Dawkins, "Preface," in *The Life and Letters of William Buckland*, ed. E. Gordon (1894); A. S. Woodward, "William Boyd Dawkins," *DNB*, supplement, *1922–30*, 250–51. Boyd Dawkins was, fittingly, the first recipient of a geology scholarship endowed by Baroness Angela Burdett-Couts, who had helped to fund the Brixham Cave excavation.

14. W. R. Chapman, "Arranging Ethnology," in *Objects and Others*, ed. G. Stocking (1985); M. Bowden, *Pitt-Rivers* (1991), 54–56.

that of the historical archaeologists. They took for granted the antiquity of the human race, the newly discovered scope of prehistory, and the existence of a pre-Celtic culture. In addition, three of the four men trained in fields other than archaeology. Evans began his career as a geologist, Lubbock as a zoologist, and Boyd Dawkins as paleontologist, and they shared a conviction that methods borrowed from geology and paleontology could be used to understand prehistory. The structure of their research program, geological archaeology, reflected their easy familiarity with the idea of a long prehistory and their backgrounds in natural science.

The geological archaeologists' use of hypotheses reflected their single greatest debt to natural science. British geologists, dealing with data known to be both fragmentary and complex, had made limited use of hypotheses for more than four decades. The average geologist seldom speculated on the causes of particular geologic features but employed hypothetical classifications and correlations of strata almost daily. Paleontologists, following the methods of Cuvier, reconstructed extinct animals by extrapolating from fossil bones and knowledge of living species. Hypotheses, in both cases, were not ends in themselves but means of organizing masses of otherwise unsuggestive facts.

The geological archaeologists faced problems that they saw as analogous to those in geology. First, they had to create a chronological framework into which prehistoric artifacts could be fitted—a framework equivalent to that provided by geology's system of eras, periods, and epochs. Once in place, such a framework would allow them to organize artifacts in roughly chronological order and to reduce the vastness of prehistory to periods of manageable size. Second, they had to develop a method of attaching meaning to the notoriously unsuggestive relics of prehistory. Without such a method, they would be unable to answer what they saw as prehistoric archaeology's crucial question: How did prehistoric Europeans live?

To create a chronological framework, Lubbock and his colleagues adopted the geologists' reliance on classificatory hypotheses. They began by adopting the Three-Age System developed by Scandinavian archaeologists, which divided prehistory into

Stone, Bronze, and Iron ages. Lubbock noted, as John Evans had in his 1859 paper on the Somme Valley, that the chipped-stone tools found in the Drift had little in common with the polished-stone tools of the Scandinavians' Stone Age. He modified the Three-Age System accordingly in his 1865 book *Pre-historic Times*. What had been the "Stone Age" became the "Neolithic," and the Drift era—with its extinct animals and chipped-stone tools—became the "Paleolithic." Lubbock's distinction between the Paleolithic and the Neolithic passed quickly into archaeological usage, making the Three-Age System into a four-age system. Subsequent hypotheses divided each of the four ages into periods, grouped apparently related artifacts into sequences, and correlated particular sequences with particular periods. Archaeologists refined, modified, and sometimes overthrew the hypotheses entirely as new data came to light. Debates over the validity of a given hypothesis were seldom resolved quickly, but the hypotheses served an important purpose. The archaeologists' classificatory schemes imposed an internal structure on prehistory, similar to that which the geologists' eras, periods, and epochs imposed on the Earth's history.

The geological archaeologists also relied on the methods and assumptions of the natural sciences in their attempts to evaluate particular subdivisions of prehistory. They borrowed, in this case, specifically from geology. The central problem in both fields was to agree upon the criteria by which questions of chronology would be settled.

British geologists' stratigraphic research program relied on their ability to trace a particular stratum through a series of separate local outcrops. Without that ability, the goal of connecting local sequences of strata into a single series would be unreachable. Members of the geological community disagreed, early in the century, on the issue of whether rock type or fossil content offered a better basis for correlating strata. By the 1850s, however, nearly all British geologists favored the use of fossils for correlation.[15] Archaeologists who confronted the newly expanded

15. A few geologists, notably Adam Sedgwick, continued to defend the correlation of strata by rock type well into midcentury.

Paleolithic faced essentially the same problem. The principle of superposition was enough to establish the order in which the artifacts at any given site had been deposited. Reconstructing a detailed chronology of the entire Paleolithic, however, meant integrating sequences of artifacts from many different sites. Integrating local sequences required, in archaeology as in geology, a basis for making correlations.

French paleontologist Edouard Lartet used animal bones as a basis for correlation. The early phase of the Paleolithic became, in Lartet's scheme, the "Hippopotamus Age." The middle and late phases became the "Mammoth/Cave Bear Age" and the "Reindeer Age." The age of a given artifact was thus determined by the animal bones with which it was associated. The stone axes from Abbeville, found among the bones of mammoths, belonged unambiguously in the "Mammoth/Cave Bear Age." Lartet's colleague, Gabriel de Mortillet, proposed an alternative classification rooted not in paleontology but in archaeology itself. De Mortillet assumed that stone tools had become steadily more sophisticated and more finely worked over time. Determining the relative ages of sites thus became a matter of comparing their stone tools—the site with the most primitive tools would be the oldest. De Mortillet eventually recognized five distinct phases in the Paleolithic, naming each for a site where its distinctive type of tool had been found (fig. 10).[16] De Mortillet's scheme went on to eclipse not only Lartet's but also a bewildering variety of modifications, compromises, alternatives, and criticisms proposed by others. Formulated in the 1860s and 1870s, de Mortillet's division of the Paleolithic remained a part of European archaeological theory well into the twentieth century.[17]

The archaeologists' debate over how to subdivide the Paleolithic, like the geologists' debate over how to correlate rock strata, concerned fundamental principles. The outcome, in each case, played a major role in determining the discipline's course for de-

16. The names—including Mousterian, Solutrean, and Aurignacian—are still part of present-day archaeological usage, but they now refer to cultures rather than to time periods.

17. G. Daniel, *150 Years of Archaeology* (1975), 99–109; B. G. Trigger, *A History of Archaeological Thought* (1989), 94–102.

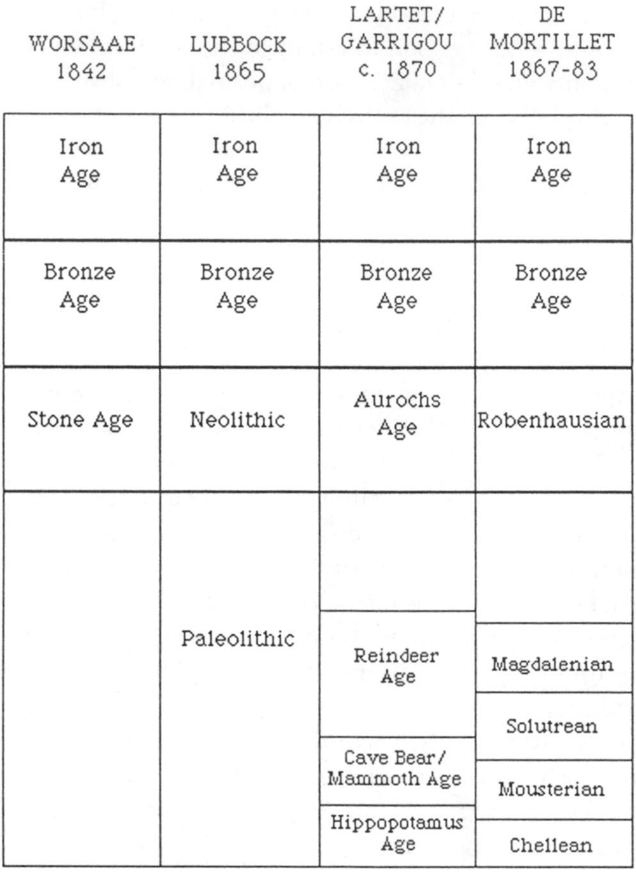

WORSAAE 1842	LUBBOCK 1865	LARTET/ GARRIGOU c. 1870	DE MORTILLET 1867-83
Iron Age	Iron Age	Iron Age	Iron Age
Bronze Age	Bronze Age	Bronze Age	Bronze Age
Stone Age	Neolithic	Aurochs Age	Robenhausian
	Paleolithic	Reindeer Age	Magdalenian
			Solutrean
		Cave Bear / Mammoth Age	Mousterian
		Hippopotamus Age	Chellean

Figure 10. Subdivisions of prehistory proposed by four European archaeologists. The names set in smaller type in the Lartet/Garrigou and De Mortillet systems refer to epochs within the Paleolithic (based on Daniel, *150 Years of Archaeology,* 1975).

cades afterward. Prestwich, Pengelly, and Falconer's investigation of human antiquity—characterized by an intense concern with the type and location of fossil bones—was shaped by the geological community's decision to correlate by fossils rather than rock type. Late nineteenth- and early twentieth-century studies of Paleolithic archaeology were, in the same way, products of de Mortillet's success. De Mortillet's use of stone-tool types as time markers mirrored geologists' use of "index fossils," reinforcing the link between geology and geological archaeology.[18]

A third area in which geological archaeologists borrowed from the natural sciences had to do with the problem of attaching meaning to the artifacts that they studied. The form of Stone Age artifacts revealed little about their function. Their uniform materials, relatively crude workmanship, and simplicity of design distanced them from anything in a nineteenth-century European's everyday experience. Determining the artifacts' original functions in the absence of written descriptions or modern analogues, had frustrated the historical archaeologists. Geological archaeologists solved the problem by drawing a parallel between artifacts and fossil bones. They believed that, just as fossil bones were the fragmentary remains of extinct animals, human artifacts were the fragmentary remains of vanished cultures. Paleontologists could draw inferences about the soft tissues of extinct animals by studying the relationship between bone and flesh in living species. Lubbock and his colleagues argued that, by analogy, archaeologists could use "modern savages" as living models for the vanished races of prehistoric Europe.

Lubbock and his colleagues borrowed from geology and paleontology in other ways besides their solutions to problems of chronology and interpretation. Their use of modern Stone Age cultures to stand in for prehistoric ones mirrored not only the paleontological analogies between living and fossil organisms but also Charles Lyell's use of contemporary geological processes as models for ancient ones. Their classification of prehistoric artifacts was strongly influenced by Linnaean taxonomy. It ignored the artifacts' geographical context, classifying them instead by their form, material, and level of workmanship. This style of classification reflected the belief—explicit in Pitt-Rivers's work, and implicit in that of both Lubbock and Evans—that human artifacts could be grouped into families, genera, species and varieties. In-

18. Historians of archaeology have often commented on the use of stone tools as archaeological "index fossils." See, e.g., Daniel, *150 Years,* 106–8; and Trigger, *Archaeological Thought,* 96.

deed, Pitt-Rivers followed Darwin in arguing that close taxonomic relationships between groups could be correlated with
close evolutionary relationships.[19]

Finally, the new archaeology followed both geology and paleontology in adopting an international outlook. Leading British
geologists had abandoned an insular view of geology early in the
nineteenth century when they set out to reduce the world's stra-
tified rocks to a generalized sequence. They recognized that Britain's rocks, or the rocks of any other country, represented only
part of that sequence. Reconstructing the stratigraphic sequence
completely meant studying rocks from a wide variety of sites in
all parts of Europe.[20] The geological archaeologists set out to solve
an analogous problem: how to construct a generalized picture of
prehistory from its artifacts. Focusing on the history of humankind rather than the history of the Earth, they collected data
at sites throughout Europe—in their own travels, by proxy, and
through correspondence with European archaeologists. In their
published work, they consistently used both British and continental material as support for their ideas.

The geological archaeologists argued explicitly for the utility
of geological methods and for what they saw as archaeology's
natural ties to geology. John Evans prefaced his 1872 book on
British stone implements with a defense of artifact-based classification schemes. He noted in a later chapter that "in seeking to
ascertain the method by which the stone implements and weapons of antiquity were fabricated, we cannot, in all probability,
follow a better guide than that which is afforded us by the manner in which instruments of similar character are produced at the
present day."[21] Boyd Dawkins's 1874 book, *Cave Hunting*, itself a
seamless blend of geology and geological archaeology, argued
that archaeological knowledge no longer depended on written

19. See Chapman, "Arranging Ethnology," 27–33; for a discussion of Pitt-
Rivers's theories of cultural evolution, see G. W. Stocking, *Victorian Anthropology*
(1987), 180–81.

20. For a discussion of international cooperation in geology, see M. J. S. Rudwick, "International Arenas of Geological Debate in the Nineteenth Century,"
Earth Science History 5 (1986); 152–58.

21. John Evans, *The Ancient Stone Implements, Weapons, and Ornaments of Great
Britain* (1872), 1–13, quotation on 13.

documents. By "the modern method of scientific research," he assured his readers, "we are able to extend the narrative away from the borders of history far back into the archaeological and geological past."[22]

It was John Lubbock, however, who emerged as geological archaeology's most enthusiastic evangelist. The opening paragraphs of *Pre-historic Times,* unchanged through three editions, stood as a declaration of independence from historical archaeology. Lubbock began where Lyell's 1859 address to the British Association for the Advancement of Science had ended, stating, "The first appearance of man in Europe dates back to a period so remote that neither history, nor even tradition, can throw any light on its origin or mode of life." He then turned to the prevalent belief that the past was "hidden from the present by a veil, which time would probably thicken, but could never remove." The image of a veiled, shrouded past was aptly—and no doubt consciously—chosen; historical archaeologists used it regularly when discussing pre-Roman sties. This belief, Lubbock noted, had led archaeologists to regard prehistoric remains as intrinsically interesting, but not as "pages of ancient history."[23] His tone left no doubt that this situation would soon change.

Having introduced the problem of understanding prehistory, Lubbock introduced his solution. "Of late years," he told his readers, "a new branch of knowledge has arisen; a new Science has, so to say, been born among us, which deals with times and events far more ancient than any of those which have yet fallen within the province of the archaeologist." A few paragraphs later, Lubbock presented the key to the "new science." There was no reason why "the methods of examination which have proved so successful in geology, should not also be used to throw light on the history of man in pre-historic times. Archaeology forms, in fact, the link between geology and history."[24] Lubbock's reference to "archaeology" in the last sentence clearly meant geological archaeology. The brief sentence offered a picture of a new intellec-

---

22. W. B. Dawkins, *Cave Hunting* (1874), 74–75.
23. J. Lubbock, *Pre-historic Times* (1865), 1.
24. Ibid., 1–2.

tual landscape: prehistory had become the subject of a new archaeological subdiscipline, called archaeology but closely allied with geology.

Lubbock then went on to explain the new archaeology's methods in more detail. He began each edition of *Pre-historic Times* by outlining the division of prehistory into Paleolithic, Neolithic, Bronze, and Iron ages. At the outset of three long chapters of ethnographic data, he expanded on his statement that archaeologists should use geological methods, noting that "the rude bone- and stone-implements of bygone ages [are] to one, what the remains of extinct animals are to the other." He emphasized the importance of studying "modern savages," drawing an explicit parallel between it and the paleontologist's study of extinct animals' living relatives. "The Van Diemaner [Tasmanian] and South American," he concluded, "are to the antiquary what the opossum and the sloth are to the geologist." [25]

Lubbock's methodological pronouncements were more than just window dressing, as a speech he delivered to the Archaeological Institute in 1866 makes clear. The occasion was symbolic—the historically oriented Institute had crated a "primeval antiquities" section—and Lubbock set out to make the most of it in his address. He reviewed each of the four ages of prehistory but devoted most of his time to a defense of the geological archaeologists' methods. He set out to satisfy the audience that such methods were "as trustworthy as those of any natural science" and carefully tailored his strategy to the interested but skeptical audience he faced. The address was designed not only to summarize "some of the principal results of modern research" but also to give the audience "some idea of the kind of evidence on which these conclusions are based." [26]

Lubbock began, as he had in *Pre-historic Times,* with a general discussion of the new archaeology's methods. He argued that techniques borrowed from natural science were responsible for "the progress recently made" in prehistoric archaeology and that such techniques "must eventually guide us to the truth" about

25. Ibid., 336–37.
26. Lubbock, "Address Delivered to the Section of 'Primeval Antiquities,'" *AJ* 23 (1867): 191.

prehistory. Brief discussions of the Paleolithic, Neolithic, Bronze, and Iron ages followed, each consisting of brief, numbered statements summarizing current archaeological knowledge and identically numbered summaries of the evidence for each point. Lubbock hoped to show, by illustrating the close connection between statement and evidence, that "archaeology . . . can arrive at definite and satisfactory conclusions, on independent grounds, without any assistance from history." His speech was less a summary of recent discoveries than a plea for acceptance of the new subdiscipline and its methods. "I care comparatively little," he concluded, "how far you accept our facts or adopt our results, if only you are convinced that our method is one that will eventually lead us to sure conclusions."[27]

Lubbock's attempts to demonstrate prehistoric archaeology's viability as a discipline focused on its ties to the natural sciences. The geological archaeologists hoped, perhaps, that by emphasizing the ideas that their discipline shared with geology and zoology, they could also share in the prestige that those sciences enjoyed. Lubbock and his colleagues said little, in writings designed to promote their new research program, about its close ties to the newly independent discipline of anthropology. The two disciplines' shared conviction that the lives of "modern savages" were similar to those of prehistoric Europeans shaped both the geological archaeologists' vision of prehistory and their attempts to understand it. The roots of the conviction lay in the emerging anthropological movement that George Stocking has dubbed "classical evolutionism."[28]

The classical evolutionists, led by Edward Tylor, sought to explain the origins of human culture without invoking the hand of God. They frequently invoked non-Darwinian mechanisms but shared Darwin's goal of linking a complex set of phenomena to a naturalistic developmental process. All societies, they argued, had the potential to develop from a state of savagery to one of civilization. Different societies would develop at different rates, and

27. Ibid., 208.
28. Stocking, *Victorian Anthropology*, chap. 5. Also see Trigger, *Archaeological Thought*; and J. W. Burrow, *Evolution and Society* (1966), chap. 7.

some might become stagnant, but all those that did develop would pass through the same sequence of stages in the same order. A particular level of toolmaking ability and particular types of social and religious institutions characterized each level, the classical evolutionists believed. Societies at the same level of development would, allowing for local variations, possess similar cultures. Modern stone-tool-using peoples in Africa or the Pacific could, therefore, be used as models for the Stone Age ancestors of modern Europeans.

This link between ancient and modern stone-tool users paid dividends to anthropologists as well as to geological archaeologists. It gave the geological archaeologists not only a means of understanding how particular artifacts might have been used but also a sense of the artifacts' broader cultural context. Geological archaeology thus sidestepped the barrier that had blocked earlier attempts to study prehistoric Europe. Modern ethnographic reports gave it a substitute for the eyewitness accounts available to students of the Roman, Saxon, and medieval eras. Anthropologists, on the other hand, interpreted the archaeologists' subdivisions of prehistory as an outline of the universal stages of cultural evolution. Each subdivision of the Stone Age was characterized, in de Mortillet's popular scheme, by a unique style of toolmaking. Taken as a sequence, they seemed to represent a natural progression from crude forms to more refined ones, in cultural terms, from savagery to sophistication.

Relations between the historical and geological branches of archaeology remained, for the most part, distant but cordial. Each accepted, though perhaps did not fully understand, the other's methods and goals. The contrast in their approaches to barrow digging—the one type of fieldwork in which both groups engaged—highlighted the intellectual gulf that separated them. Historical archaeologists who participated in barrow digging saw it as a means of fact gathering. They sought to amass a detailed, carefully documented record of a particular class of antiquities, and some of the most ambitious devoted a lifetime to work in a single country.[29] The geological archaeologists who excavated

29. J. R. Mortimer, *Forty Years Digging in the Celtic and Saxon Grave-Hills of Yorkshire* (1905).

barrows were not fact-gathering specialists. They saw excavation as a means to the end of formulating and testing theories about the lives of prehistoric Europeans and barrows as one among many types of excavation site. Books such as Lubbock's *Prehistoric Times* discussed British barrows but gave equal or greater attention to Swiss lake dwellings, French caves, and Danish refuse heaps.

A few historical archaeologists refused to accept their new counterparts as fellow scientists and fellow investigators of the past. They remained skeptical of the new geological archaeologists' reliance on hypotheses and on methods drawn from natural science. Thomas Wright, a founder of historical archaeology and vocal critic of geological archaeology, presented his case at the 1866 Birmingham meeting of the British Association for the Advancement of Science. Wright agreed that Lubbock and his colleagues had created a new style of archaeology; that, he argued, was precisely the problem. Rather than treat the historical archaeologist as an equal partner, Wright complained, the geologist "has created a sort of archaeology of his own, made in the first place to suit his own theories, and he takes only the advice of those which will give him an opinion which is in accordance with a foregone conclusion, and this is often quite contrary to the teachings of archaeological science. Archaeology, as a science, has now reached too high a position to be treated with so little respect."[30]

The detailed criticisms of Lubbock's work that followed stemmed from Wright's commitment to the strict empirical methods of historical archaeology. Wright first accused Lubbock of uncritically accepting the Three-Age System—a "mere delusion" sprung from "too hasty generalizing" and too little hard data. He then charged that Lubbock's commitment to the Three-Age System rendered his observations suspect. Wright listed examples of stone implements found with metal ones, and of bronze swords—assigned by Lubbock to the pre-Roman Bronze Age— found with Roman artifacts. These examples, he argued, showed that stone, bronze, and iron implements had all been used simultaneously in Britain, around the time of the Roman invasion.

30. T. Wright, *The Celt, the Roman, and the Saxon,* 3d ed. (1875), 2.

Lubbock had chosen to ignore them rather than relinquish his cherished a priori notion of sequential Stone, Bronze, and Iron ages.

Neither Lubbock nor Wright relented, and the dispute between them remained unresolved. Their positions were too deeply rooted in distinct, incompatible views on how to study the past. Lubbock answered Wright's 1866 criticisms in the 1869 edition of *Prehistoric Times* (the title had now dropped its hyphen), and Wright offered a rebuttal in the 1875 edition of *The Celt, the Roman, and the Saxon*, but these were little more than restatements of earlier positions. Wright remained committed to the belief that a single contrary fact was sufficient to destroy the most complex theory. Lubbock, following a different scientific tradition, saw theories such as the Three-Age System as an important tool for organizing data. He regarded Wright's counterexamples as isolated anomalies that did not constitute a serious challenge to a well-established theory.

Not surprisingly, scientists outside the archaeological community took Lubbock's point of view. Geological archaeology's close philosophical and methodological alliances with geology on the one hand and anthropology on the other helped to ensure acceptance of the new research program. The same ties helped to ensure that, after the mid-1860s, members of all three communities took an avid interest in studies of prehistoric Europe.

The Grand Alliance: Geology and Geological Archaeology

Until 1859, the intellectual boundary between geology and archaeology coincided with the stratigraphic boundary between the Post-Pliocene and Recent epochs. The two sciences shared broadly similar goals—understanding and reconstructing the past—but studied different data using different methods. The establishment of human antiquity meant that geology and a newly expanded archaeology overlapped for the first time. The common ground that geology and archaeology shared remained relatively narrow, however. Prehistoric archaeologists studied all periods of European prehistory, from the time of the earliest Paleolithic hunters to that of the bronze- and iron-using tribes conquered by the Romans. Geologists' concern with humankind was limited to the Paleolithic Age, when humans had coexisted with now-extinct animals in a climate very different from that of present-day Europe. Much of the prehistoric archaeologists' intellectual domain, and all of the historical archaeologist's, lay outside the geologist's area of interest (fig. 11). Joseph Prestwich expressed a belief well established among geologists when, in 1891, he described the Neolithic Age as "a subject for the archaeologist," and "geologically unimportant."[31]

The simultaneous emergence of classical evolutionism and geological archaeology created a new, intimate relationship between anthropological and archaeological studies of prehistory. The relationship heightened anthropologists' interest in the emerging picture of prehistoric Europe pieced together by the archaeologists. Anthropologists' interest did not, however, extend to the details of individual sites or small groups of artifacts. Their principal concern was the larger question of how European culture had developed—what intermediate steps separated the stone tools of the Drift from the iron swords with which Celtic warriors had fought the Romans. The detailed investigation of prehistoric sites thus remained the province of archaeologists and, in the case of Paleolithic sites, of geologists as well. Collaboration between the two disciplines was important but limited. Both studied Paleolithic sites and collected similar types of data, but they asked different questions of the data and sought different types of answers.

The British geological community's interest in the age of the human race did not end with the establishment of human antiquity. The core set that investigated Brixham Cave and the Somme Valley dissolved by 1863, but individual geologists—Prestwich, Pengelly, and James Geikie, for example—continued to study the human antiquity problem. Their goal was a more precise picture of when, during the Post-Pliocene, the first humans appeared in Europe. Their research focused on the stratigraphic and paleontological context of early Paleolithic artifacts. The study of early humans remained, for them, part of the larger geological problem

31. J. Prestwich, "On the Age, Formation, and Successive Drift-Stages of the Valley of the Darent," *QJGSL* 47 (1891): 156.

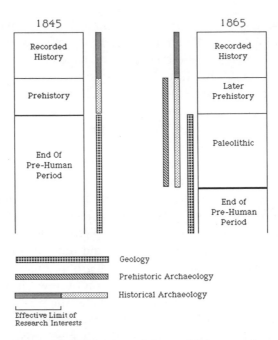

Figure 11. The impact of the establishment of human antiquity on the disciplinary boundaries of British geology and archaeology. The two-tone shading used to represent historical archaeology reflects the difference between historical archaeologists' stated and actual research interests.

of understanding the vanished world of the Post-Pliocene. They were concerned primarily with the human species as a species— as one component of the Post-Pliocene's characteristic fauna.

The geological archaeologists saw Paleolithic humans in a different context and sought answers to different questions. They studied Paleolithic humans not as one among many species in a complex fauna, but as a species made unique by its possession of culture. Their research focused on the material portions of that culture, on its regional variations, and on how it gave way to the more advanced cultures of the Neolithic and Bronze ages.[32]

32. Nineteenth-century scientists spoke of "human culture" in the abstract, but not of "human cultures" as specific, distinct entities. I have deliberately used the anachronistic plural form in this chapter for the sake of clarity.

Geological archaeologists were less concerned than British geologists were with when Paleolithic humans had lived. They were less concerned than French archaeologists with how to subdivide the Paleolithic into periods or epochs. They took an interest in both issues, but their own research focused on the question, "How had prehistoric man lived?"

John Evans's second paper on flint tools from the Somme Valley expressed this distinction concisely. In it, Evans described the age of such tools as a matter for geologists and their form as a matter for archaeologists. Correlating the two, he noted, would require close interdisciplinary cooperation.[33] The many works on Paleolithic humans published in Britain after 1860 consistently reflected the same distinction. Geologists, even when they published in archaeological and anthropological journals, dealt exclusively with chronological issues.[34] Geological archaeologists, though usually well acquainted with current discussions of human antiquity, used them only as background for discussions of Paleolithic culture. The approaches taken by Lyell and Geikie, on one hand, and Lubbock and Evans, on the other, illustrate this difference.

The first eleven chapters of Lyell's *Antiquity of Man* (1863) covered familiar intellectual ground. They argued for the presence of humans in the Post-Pliocene by presenting the wide range of evidence that had been collected during the preceding five years. In subsequent chapters, however, Lyell took human antiquity for granted and turned to two related subjects: Darwin's theory of evolution and the glacial period and its chronological relationship to the first appearance of humans in Europe. Lyell's discussion of humans and the glacial period focused on a question new to the human antiquity investigations. Five years of concentrated effort had shown that humans had lived during the Post-Pliocene; in chapter 12 of *Antiquity,* Lyell asked when, during that epoch, they had appeared.

Though many contemporary writers regarded Lyell's chapters

33. John Evans, "Account of Some Further Discoveries of Flint Implements," *Arch.* 39 (1862): 80–81.

34. See, e.g., Joseph Prestwich's paper, "On the Primitive Characters of the Flint Implements of the Chalk Plateau of Kent," *Journal of the Royal Anthropological Institute* 21 (1892): 246–62.

on glacial geology as a long digression, Lyell saw them as an integral part of his discussion of human antiquity.[35] Eleven chapters of evidence for human antiquity preceded the seven "glacial" chapters, and a recapitulation of the evidence followed them. Five of the seven glacial chapters refer directly to humankind in their titles. Finally, the chapters relating the human period to the glacial period are consistent both in method and objective with the earlier chapters on human antiquity. In both sections of the book, Lyell used stratigraphic and paleontological techniques to determine humankind's proper place in the geologic record. The first eleven chapters place the human race in a particular geological epoch; the next seven locate it within that epoch.

Antiquity of Man was not, as some historians have suggested, a treatise on anthropology written by a geologist. It was, rather, a treatise on Post-Pliocene geology written by a geologist. Chapters 12–16 inquired into "whether the peopling of Europe by the human race and by . . . mammalia now extinct" had occurred while glaciers still filled most mountain valleys.[36] Lyell made it clear, early in chapter 13, that the glacial chapters were part of a broad program of geological research. "I know of no inquiry better fitted to clear up our views respecting the geological state of the northern hemisphere at the time when the fabricators of the flint implements of the Amiens type flourished. I shall therefore now proceed to consider the chronological relations of that ancient people with the final retreat of the glaciers."[37] Like his fellow geologists, Lyell found the world that the first Europeans inhabited more interesting than the world that they made.

James Geikie's *The Great Ice Age*, published eleven years after *Antiquity of Man*, offered readers a new picture of early humans and their geological context.[38] Geikie's work incorporated new stratigraphic data and an interpretation of the glacial period that emphasized continentwide ice sheets, but its treatment of Stone Age Europeans was similar to Lyell's. Its discussion of humankind

35. W. F. Bynum, "Charles Lyell's *Antiquity of Man* and Its Critics," *Journal of the History of Biology* 17 (1984): 183–85.

36. C. Lyell, *Geological Evidences of the Antiquity of Man* (1863), 229–30.

37. Ibid., 230.

38. J. Geikie, *The Great Ice Age* (1872), chaps. 29–32.

was tied closely to geological questions about Europe's Post-Pliocene climate and fauna. Geikie's particular concern was the distribution of Paleolithic artifacts and animal remains in Britain. Why, he asked, is Southeast England the only area where either is abundant—or found at all, outside of caves—when Neolithic bones and tools are found throughout the British Isles?[39]

Geikie explained the distribution of Paleolithic artifacts the same way he had explained the mixture of tropical and arctic animals found in British Post-Pliocene deposits. The Post-Pliocene, he argued, had seen alternating cold and warm—glacial and interglacial—periods. Paleolithic humans had flourished throughout England during one of the warm periods, alongside tropical or temperate faunas. During a subsequent cold period, advancing ice sheets and rising seas had driven Paleolithic humans and animals alike out of Britain, obliterating their remains everywhere but in caves and the comparatively warm Southeast. The Neolithic began much later, as the climate warmed, the glaciers and seas retreated, and a new race of humans migrated into Britain across what is now the English Channel.[40] European and North American deposits confirmed this theory, he noted; Paleolithic remains appeared only in interglacial deposits, and a sharp stratigraphic break separated them from the Neolithic remains above.[41]

The questions that geologists asked about Post-Pliocene humans after 1863 resembled those that Falconer had asked about Post-Pliocene elephants and rhinoceroses in the 1850s: When, geologically, did they first appear in Europe? How extensive was their range? With what animals did they coexist, and when? Questions about the stratigraphic position, faunal association, and geographic distribution of human remains were part of a single program of geological research. British geologists studied Paleolithic artifacts after 1863 in much the same way that they had done before—as one aspect of an important but imperfectly understood period in the Earth's history. They were not concerned,

39. Ibid., 440–51.
40. Ibid., 448–51.
41. Ibid., 452–68.

except in the most general terms, with the people who had created those artifacts.

Understanding the "native behind the artifact" was, in contrast, the geological archaeologists' primary concern. The heart of Lubbock's *Pre-historic Times* comprised seven chapters—some of them originally separate papers—that described the artifacts characteristic of particular European regions and particular prehistoric periods. One described the Bronze Age dwellings found along the old shoreline of Switzerland's Lake Constance; another analyzed the refuse heaps left behind by Danish fishermen in the Neolithic; a third examined the Paleolithic cave dwellers of southern France.[42] Brief references throughout the book, along with three complete chapters at the end, used comparisons with "modern savages" to illuminate aspects of prehistoric life not preserved in the archaeological record. Lubbock took the quality of life in prehistoric Europe as his theme; each of his central chapters sketched one stage in what he saw as humankind's ignoble, barbaric past.[43]

Lubbock discussed human antiquity, but only as a way of demonstrating the scope of prehistory. He discussed the extinct animals that coexisted with Paleolithic humans, but only as a way of showing how different the Paleolithic world was from the modern one.[44] He described the changing climate of late Post-Pliocene Europe, but only as a way of establishing the great antiquity of the Paleolithic Age or illustrating the harsh conditions that prehistoric humans endured.[45] Lubbock incorporated the facts of Post-Pliocene geology into his discussion of prehistory as background, rather than as an independent theme. His descriptions of the Post-Pliocene landscape were simply stage settings for the drama of human prehistory.

42. Lubbock, *Pre-historic Times,* chaps. 6–8.

43. Stocking, *Victorian Anthropology,* 151–56. Lubbock's second book, *The Origins of Civilization* (1870), presented a much more optimistic view of prehistoric humans' mental capabilities and offered an explicitly evolutionist interpretation of human culture. Lubbock revised his picture of the past, Stocking argues, in response to the degenerationist theories proposed by the Duke of Argyll and Archbishop Whatley.

44. Lubbock, *Pre-historic Times,* 237–43; subsequent editions treated Europe's Post-Pliocene fauna in a separate chapter.

45. Ibid., 300–301, 390–406.

John Evans's 1872 book, *The Ancient Stone Implements, Weapons, and Ornaments of Great Britain*, embodied an approach to prehistory very similar to the one Lubbock used in *Pre-historic Times*. Despite its title, which suggests a dry catalog of facts in the style of the historical archaeologists, *Ancient Stone Implements* offered a reconstruction of selected aspects of Stone Age life. Evans's classification of Neolithic stone tools relied as much on function as on form. It depended, therefore, on a sparing but continual use of hypotheses about prehistoric European cultures. Each of Evans's nineteen chapters on the Neolithic discussed a particular group of implements; they bore titles such as "Perforated and Grooved Hammers," "Grinding Stones and Whetstones," and "Borers, Awls, or Drills." Each chapter cataloged variations in form and material but also discussed the uses of the implements in question. Evans described not only the shapes of Neolithic axes but also the methods for hafting them; not only the principal types of scrapers but also their use in dressing hides and starting fires.

Ethnographic data figured less prominently in *Ancient Stone Implements* than in *Pre-historic Times* but were no less important. Within the context of particular chapters, Evans frequently discussed modern stone age cultures in order to show how a particular class of implements might have been made or used. His theory that flint scrapers also served as fire starters was bolstered, for example, by a discussion of such practices among Eskimos. His chapter on hammer stones and food preparation included observation of tribes in North and South America, Australia, and Africa.[46] Evans cited hundreds of observations by dozens of ethnographers, but he used this ethnographic data for a limited set of purposes. He focused less on the workings of Stone Age Europeans' minds than on the works of their hands. *Ancient Stone Implements* was not an attempt to completely reconstruct a vanished culture, but was a survey of ancient technological ingenuity as applied to the problems of everyday life.

The geologists' attempts to define more precisely the age of the human race complemented the geological archaeologists' attempts to reconstruct Paleolithic culture. The two groups shared

46. Evans, *Ancient Stone Implements*, 274, 224–26.

data regularly, but their research agendas were different enough that disputes over priority of discovery rarely occurred.[47] Each group regularly incorporated the other's conclusions into its work—Lyell, for example, cited Lubbock's papers on Neolithic and Bronze Age life, and Evans discussed the latest estimates of human antiquity. Though they studied the same subject, the two communities shared philosophical assumptions so similar that disputes over method and technique never arose. In a period when geology's relations with physics grew increasingly turbulent, its relations with prehistoric archaeology remained consistently placid.[48]

The absence of interdisciplinary strife did not, however, ensure the absence of controversy. A debate over the existence of "Tertiary Man" waxed and waned in Britain from the late 1870s until the early 1910s. At issue were the status of what French archaeologist Gabriel de Mortillet called "eoliths." Defenders of the Tertiary Man theory interpreted the "dawn stones" as crude stone tools made by the barely human inhabitants of Pliocene (or even Miocene) Europe. Critics insisted that the stones bore no marks of human workmanship, and believed that they were nothing more than accidents of nature. Each side counted both archaeologists and geologists among its members. Prestwich, Geikie, and Pitt-Rivers found the evidence for Tertiary Man convincing, while Evans and Boyd Dawkins opposed it.[49] The fact that both geologists and archaeologists cooperated on both sides of the eolith controversy reflects the changes in disciplinary structure wrought in the 1860s by the establishment of human antiquity.

The excavation of Kent's Cavern, which began in 1865, also

47. Evans's correspondence with Lyell was particularly long and detailed; Evans appears to have been Lyell's chief source of information on Paleolithic research after 1864 (see Lyell Papers, EUL).

48. Controversy over the age of the Earth—and over the proper relationship of geology to physics—reached a peak in the late 1860s as T. H. Huxley attempted to defend geology's methods and conclusions against the criticisms raised by William Thomson (later Lord Kelvin); see J. D. Burchfield, *Lord Kelvin and the Age of the Earth* (1975).

49. See F. Spencer, "Prologue to a Scientific Forgery," in *Bones, Bodies, Behavior,* ed. G. Stocking (1988); and F. Spencer, *Piltdown* (1990), 1–28.

reflected the extent of cooperation between geologists and archaeologists.[50] Unlike the earlier dig at nearby Brixham Cave, the Kent's Cavern excavation involved scientists from both disciplines. Sponsored by the British Association for the Advancement of Science, rather than the Geological Society, it was directed by a committee that included geological archaeologists—Lubbock, Evans, and later Boyd Dawkins—as well as leading geologists. The committee, which included Pengelly and other veterans of Brixham Cave, adopted methods based directly on those Pengelly had devised in 1858. Excavators removed the cave deposits completely and systematically. The upper deposits were dug out layer by layer, but the thick bed of "cave earth" was removed a foot at a time, in one-by-three foot sections. The scientist overseeing the excavation placed the contents of each section in a separate, numbered box, recorded the contents in a log book, and then carried the specimens away to be cleaned and labeled.

Kent's Cavern, considerably larger than nearby Brixham Cave, held out the promise of a larger—and so more representative—sample of early human artifacts and Post-Pliocene animals. As a result, a detailed study of the cave appealed to geologists and geological archaeologists alike. On one hand, such a study would provide a more detailed picture of Britain's changing Post-Pliocene fauna and of humankind's relationship to it; on the other, it would better illustrate the range of tools that early Britons made. The excavation, to which the British Association gave eighteen hundred pounds and Pengelly fifteen years of his life, fulfilled expectations many times over. It produced staggering quantities of data: tens of thousands of animal bones, and artifacts of chipped stone, polished stone, bone, bronze, and clay.[51] Boyd Dawkins's *Cave Hunting* and *Early Man in Britain*,[52] Evans's *Ancient Stone Implements*, Pengelly's many lectures on human antiquity, and the post-1870 editions of *Pre-historic Times* and *Antiq-*

50. For a compact, detailed discussion of the Kent's Cavern excavation, see T. G. Bonney, "The Scientific Work of William Pengelly, FRS," in *Memoir of William Pengelly*, ed. H. Pengelly (1897).

51. Ibid., 311–13.

52. W. B. Dawkins, *Early Man in Britain* (1880).

uity of Man all discussed the results of the Kent's Cavern excavations.[53]

The Kent's Cavern project spanned fifteen years in which British geology and archaeology changed significantly. The excavations began when the consensus on human antiquity formed in 1859 was still new; they ended after the consensus had been assimilated by an entire generation of scientists. In 1865, prehistoric archaeology had only recently emerged as a distinct subdiscipline; in 1880, it had not only matured but had come to represent the cutting edge of British archaeological research.

Heterodoxy and Hegemony

Archaeology and the study of prehistory have often been seen as virtually identical. Archaeological studies of historical periods and the role of other disciplines in the study of prehistory have received little attention from historians. Standard accounts of the development of British archaeology assume that neither remained significant after the emergence of prehistoric archaeology around 1860. Such assumptions are unfounded. Geologists played an important, ongoing role in the study of Paleolithic humans after 1863. Historical archaeology remained popular, both as a subdiscipline and as a research tradition, well into 1880s and beyond. To what, then, can prehistoric archaeology's high profile be attributed? Why have historians so long assumed that, after roughly 1860, British archaeology focused exclusively on prehistory and studies of prehistory were exclusively archaeological?

The answer to these questions is twofold. It involves, first, a series of changes that decreased the visibility both of the historical archaeologists and of the geologists who studied Paleolithic Age Europe. It also involves the geological archaeologists' success in integrating themselves and their research tradition into the Victorian scientific community.

Historical archaeology remained alive and well in the decades after the establishment of human antiquity. It rested, however, in an increasingly precarious disciplinary niche between history and

53. Examples include W. Pengelly, *Kent's Cavern: Its Testimony to the Antiquity of Man* (1876); and W. Pengelly, *Report on Two Lectures . . . on the Ancient Cave Men of Devon* (1870).

geological archaeology. Philippa Levine has argued persuasively that, because the historical archaeologists were amateurs, they were excluded from Britain's newly professionalized historical community in the 1870s and 1880s.[54] The historical archaeologists saw themselves as humanists and frequently defined their science as an adjunct to, if not a servant of, history. Their self-proclaimed goal was to collect facts that historians would weave into narratives, the accuracy of which historical archaeologists would monitor. The historians' repudiation of this relationship deprived historical archaeology of its principal reason for existing. Historical archaeologists could still create their own interpretations of the past, but professional historians were likely to accord the interpretations only second-class status. One result of professional historians' nearly total rejection of historical archaeologists' work on the Middle Ages was that, by the early twentieth century, "there was no medieval archaeology" in Britain. Professional historians who expressed an interest in archaeology were considered "abnormal, possibly eccentric or perverse," and archaeologists interested in the Middle Ages were obliged to learn the science by working on Roman-era sites.[55]

Historical archaeology's relationship to geological archaeology remained amicable but disinterested. Distinct subject matter, aims, and methods created a deep gulf between the two subdisciplines. Historical archaeologists' principal concern had always been local: the history of a particular town or country as it related to the development of the British nation. That local focus remained one of the roots of historical archaeology's appeal. Geological archaeologists, on the other hand, studied a period in which national boundaries did not exist but cultural boundaries were all-important. Their theories, like the stratigraphic systems of geologists, were attempts to find patterns in data from all parts of Europe. Historical archaeology traced the history of a single people, geological archaeology, that of all Europeans.

Geological archaeology's focus on pan-European prehistory also diminished the role that amateurs could play in the study

54. P. J. A. Levine, *The Amateur and the Professional* (1986), chaps. 5–7, esp. pp. 170–76.

55. C. A. R. Radford, quoted in Hudson, *Social History of Archaeology,* 127–29.

of prehistory. Geological archaeology could not be pursued, as historical archaeology could, close to home or in spare evening hours. Prehistoric artifacts, unlike the facts that historical archaeologists collected, were unlikely to be found in churchyards and census books. Discovering the artifacts required arduous searches and, often, extensive excavation. Explaining their function and linking them to a particular people depended on hypotheses, rather than on careful induction.

Cut off from historians for professional reasons and geological archaeologists for methodological ones, the historical archaeologists became increasingly isolated. The boundaries of their subject matter had been sharply curtailed. The disciplinary landscape that emerged in the 1860s had a place neither for their studies of British prehistory (a minor part of their science) nor for their work on the medieval and Tudor periods (a major part). Historical archaeologists continued to do important work on Roman and Saxon sites, but their dreams of a partnership with historians and a comprehensive history of England based on "true Baconian principles" faded.[56]

Geological studies of human antiquity continued after 1860. Their visibility diminished, like that of historical archaeology, but for different reasons. Geologists who worked on the human antiquity problem after 1865 concentrated exclusively on questions of relative age—the relationship of early humans to the glacial period and the possible presence of humans in the Tertiary period. Though important, the answers to these questions involved revisions in Victorians' existing understanding of the past rather than forming the basis of a new one. They refined, rather than replaced, the view of human antiquity that geologists had forged in 1858–63. A clear solution to the problem of absolute age might have wrought greater changes, but such a solution was no closer in the 1870s and 1880s than it had been earlier.

The geologists most concerned with the human antiquity

56. The history of Roman and Saxon archaeology in Britain has yet to be written. Readers interested in the historical archaeologists' contributions might begin, however, by examining the nineteenth-century work of Thomas Wright and Charles Roach Smith on Roman sites and that of J. Y. Akerman and John Mortimer on Saxon remains.

question had also become less professionally visible by the late 1870s. The human antiquity core set had long since dissolved, and many of its members had ceased to play an active role in the geological community. Evans left geology for prehistoric archaeology in the early 1860s. Hugh Falconer died in 1865, and Lyell in 1875. Robert Godwin-Austen retired from geology in 1877, and Pengelly—exhausted, perhaps, by fifteen years of work in Kent's Cavern—did the same in 1880.[57] The most astute geological commentators on the new case for human antiquity were also gone by the late 1870s. Henry Darwin Rogers died in 1866, James Forbes in 1868, and John Phillips in 1874. Research on human antiquity was dominated, in the 1870s and after, by younger geologists who saw human antiquity as a geological fact rather than a stunning revelation.

Broad changes in the social and intellectual structure of British geology helped to make the human antiquity problem a less prominent part of the discipline after 1865. The Post-Pliocene was the last major subdivision of the geologic time scale to be carefully investigated. The redefinition of its upper boundary by the human antiquity core set was the last major change to be made in the geologic time scale. Adam Sedgwick's bitter dispute with Roderick Murchison over the proper classification of the oldest rocks had already ended in victory for Murchison. A great deal of stratigraphic work remained to be done on the rocks of particular periods and epochs, but the glory days of carving out and naming entire divisions of the geologic record were over. "Nothing is better settled," John Phillips announced in 1864, "than the series of great events in our geologic history."[58] The settling of the Post-Pliocene's boundaries had closed the stratigraphic frontier.

The cutting edge of British geological research shifted, during the last decades of the nineteenth century, into new fields: geochemistry, geophysics, and studies of mountains and glaciers. Young geologists such as Murchison's protégé Archibald Geikie,

57. H. B. Woodward, "Robert Alfred Cloyne Godwin-Austen," *Geological Magazine,* n.s. 2 (1885): 8; H. Pengelly, *Memoir of William Pengelly,* 266.
58. J. Phillips, "Presidential Address to Section C," *BAAS Report* 34 (1864): 45.

though still well acquainted with stratigraphy and paleontology, made their scientific reputations elsewhere. Some senior members of the geological community changed their specialties in midcareer. Joseph Prestwich turned, in the last stages of his career, from stratigraphy to hydrology. Appointed to fill the professorship of geology at Oxford in 1875, he devoted much of his inaugural lecture to geochemistry and the nature of the ice age.[59]

British geologists' intellectual horizons broadened during the 1860s and 1870s, but at the expense of the single-minded sense of purpose that had unified the geological community in the 1830s and 1840s. The age of the oldest humans remained a stratigraphic problem throughout the nineteenth century and well into the twentieth. After 1865, however, it became one among many problems for specialists in one branch of geology to pursue. It was no longer a major stratigraphic issue, and stratigraphy was no longer the alpha and the omega of British geological research. The geological aspects of human antiquity faded, as a result, from the thoughts of all but the few who had a specialized interest in them.

Prehistoric archaeology, on the other hand, seldom strayed far from the spotlight during the late nineteenth century. It dealt in artifacts less spectacular than those uncovered by classical, Egyptian, or Near Eastern archaeologists but was inextricably linked to major scientific problems. The geological archaeologists' focus on prehistoric Europe and close methodological ties to anthropology made their work inseparable from that of the classical evolutionists. The material remains of European prehistory could, many believed, serve as a template for reconstructing the cultural development of all human societies. The fact that many theorists saw cultural evolution as a reflection of physical development tied archaeology, albeit somewhat loosely, to debates over humankind's alleged apish ancestry. It is, perhaps, no coincidence that John Lubbock, the most prolific of the geological archaeologists, was also a leading proponent of cultural evolution and a thoroughgoing Darwinist.

The artifacts left behind by prehistoric Europeans were not,

---

59. J. Prestwich, "The Past and Future of Geology," *Smithsonian Annual Report for 1875* (1876), 175–95.

in themselves, startling. The interpretations attached to those artifacts and the reconstructions of early human history that rested
on them were more surprising. Those interpretations and reconstructions raised a variety of important questions with philosophical and religious, as well as scientific, implications. Had humans
evolved from a lower form of life? Had the first humans been
mere savages? If so, when and how had they acquired culture?
When, and from what source, had humans acquired moral sense
and the ability to reason? Most fundamentally, when did humankind become human?[60] Nonspecialists had ample reason,
throughout the late nineteenth century, to be interested in archaeologists' work on the first Europeans.

The geological archaeologists had close intellectual ties with
both the natural and the social sciences and used them to form
institutional ties. They had no society of their own but found out-
lets for their views in several existing societies. They did not form
a distinct community in the 1860s and 70s, but their connections
with Britain's scientific elite helped to ensure recognition of their
ideas in established disciplinary communities. Lubbock, Evans,
and Boyd Dawkins were all members of the Geological Society
by 1860, and Evans served as secretary in 1866–74 and president
in 1874–6. Elected to the Royal Society in 1857 at the astonish-
ingly young age of twenty-three, Lubbock served three terms on
its council and two as a vice-president between 1860 and 1880.
The geological archaeologists were also well represented at the Society of Antiquaries: Evans had been a member since the 1850s,
and Lubbock and Pitt-Rivers were elected in 1864. As a group,
they could present their ideas to more scientists in more disciplines than the comparatively insular historical archaeologists.

The Ethnological Society of London, however, was the geological archaeologists' most important organizational base. Evans,
Lubbock, and Pitt-Rivers were all members by 1861.[61] Lubbock's

60. General discussions of these themes include: Stocking, *Victorian Anthropology;* Burrow, *Evolution and Society;* Bowler, *Invention of Progress;* and N. R. Gillespie, "The Duke of Argyll, Evolutionary Anthropology, and the Art of Scientific
Controversy," *Isis* 68 (1977): 40–54.

61. Stocking, *Victorian Anthropology,* 247–49; E. Richards, "Huxley and Woman's Place in Science," in *History, Humanity, and Evolution,* ed. J. Moore (1989),
261–70.

election to the presidency in 1863 set the stage for their "capture," or perhaps colonization, of the venerable but hidebound society. T. H. Huxley and anatomist George Busk, evolutionists and key figures in the debate over human origins, joined the society in 1863. Alfred Russel Wallace (1866), Edward Tylor (1867), and Joseph Hooker (1868) followed them in subsequent years. The arrival of the "Lubbock Circle" changed both the power structure and the scientific focus of the Ethnological Society. It became, in the late 1860s, a forum for discussions of Darwinism, human cultural evolution, and prehistoric archaeology. Lubbock and his colleagues set out to create an alternative to the anti-Darwinian Anthropological Society of London, and they succeeded. The 1871 merger of the two societies into the evolutionist, prehistorically oriented Anthropological Institute completed the Lubbock Circle's triumph.[62]

Personal connections played an equally important role in the success of geological archaeology. Evans, of course, knew leading geologists such as Prestwich and Lyell from his work on the human antiquity problem. Boyd Dawkins had a growing reputation within the geological community, in addition to his notoriety as a promising archaeologist. Lubbock, the unofficial leader of the geological archaeology movement, was also its best-connected member. Befriended at the age of eight by his family's neighbor, Charles Darwin, he became part of the Darwinian inner circle while still in his early twenties. His friendship with Darwin led to friendships with many of the young professionals who shaped late Victorian science, T. H. Huxley, Joseph Hooker, John Tyndall, and George Busk among them. Lubbock also became a member of Huxley's influential X-Club, and when Huxley's circle founded the *Natural History Review* as a journal for "young men with plastic minds," he was one of the editors.[63] Evans and Pitt-Rivers worked

62. D. K. Van Keuren, *Human Science in Victorian Britain* (1982), 18–77. Recent discussions of the "Lubbock Circle" and its institutional foundations include W. R. Chapman, "The Organizational Context in the History of Archaeology," *Antiquaries' Journal* 69 (1989): 23–42; and D. K. Van Keuren, "From Natural History to Social Science," in *The Estate of Social Knowledge*, ed. J. Brown and D. Van Keuren (1991).

63. H. G. Hutchinson, *The Life and Letters of Sir John Lubbock*, 1:23–45; Stocking, *Victorian Anthropology*, 151–53.

with many of the same rising scientific stars—as well as with established figures such as Francis Galton and A. W. Franks of the British Museum—at the newly remade Ethnological Society.

The geological archaeologists' personal connections did not stop at the English Channel. Lubbock and his cohorts shared data, made field trips, and conducted excavations with leading Continental archaeologists. Henry Christy, for example, collaborated with Edouard Lartet on his investigations of cave dwellings in the Dordogne region of France. Lubbock studied ancient Danish shell middens with Japetus Steenstrup and Swiss lake-edge dwellings with Adolphe Morlot. Evans, Lubbock, and an international party of archaeologists visited Switzerland in 1869 to excavate the large Iron Age cemetery at Hallstatt.[64] These collaborations, based on shared assumptions about goals and methods, brought the geological archaeologists and their work international prominence.[65] It also made them part of an international scientific community from which the historical archaeologists were isolated. Geological archaeologists appeared in force at the first international Congress of Archaeology and Anthropology, held in Paris in 1867. The 1868 congress took place in London and Norwich, with Lubbock serving as president and Pitt-Rivers as secretary.

Prehistoric archaeology became a distinct subdiscipline in Britain during the early 1860s. It was well established by 1870, buoyed both by a complementary intellectual climate and by the concerted efforts of its first practitioners. Well integrated into the British and Continental scientific communities, Lubbock and the geological archaeologists were in an ideal position to put their new approach to the past into practice. They stood, in effect, where British geologists had stood fifty years earlier. A vast span of imperfectly understood time stretched before them; a growing body of data demanded organization and interpretation; and although they had done significant work, dozens of important questions remained unexplored. Geological archaeology's own "golden age" began in the 1870s and, from the beginning, investigations of the first Europeans were an integral part of it.

64. Daniel, *150 Years*, 89–111; Keith, "Anthropology," 70–79.
65. For one European's discussion of these shared assumptions, see A. Morlot, "Introductory Lecture to the Study of High Antiquity," *Smithsonian Annual Report for 1862* (1863): 303–15.

EIGHT

A New Way of Seeing

GEOLOGISTS had discussed the age of the human race many times before John Philp broke through the roof of Brixham Cave in January 1858. The presence of men among the mammoths had been proposed many times before William Pengelly and Henry Keeping found stone tools beneath the cave's bone-studded floor. Pengelly himself had suggested the idea in the 1840s on the strength of his work in nearby Kent's Cavern. The discoveries of Pengelly, Schmerling, Tournal, and others ensured that, well before Joseph Prestwich and John Evans returned from France in 1859, British geologists knew both the details and the potential implications of the human antiquity question. The writings of theorists such as Lyell, Mantell, and Phillips suggest that, even before 1859, the geological community would have accepted human antiquity if presented with compelling evidence.

The events of 1858–63 established the idea that humans had existed in the Post-Pliocene, and made that idea part of orthodox science. One contemporary observer, writing in 1868, dubbed those events a "great and sudden revolution."[1] It is accurate, and in many cases useful, to see the revolution of 1858–63 as an end point—the climax of a sixty-year series of attempts to establish human antiquity. To fully understand the impact of human antiquity on Victorian thought, however, it is essential to take a broader view. The revolution, though significant, was only one episode in a long and complex story. It changed Victorian scientists' picture of the past, but—in the years after 1863—it also changed the ways in which they studied the past.

1. C. Murchison, Editor's Note, in *Paleontological Memoirs and Notes of High Falconer* (1868), 2:486.

Victorian scientists, clergymen, and laypersons held views of the past based on a shared group of interrelated ideas. Different groups and individuals elaborated on the foundation differently but nonetheless shared its basic assumptions. Natural theology, the belief that the structure of the Earth and the history of life provided evidence of God's existence and benevolence, was one important element. Organic progression—the idea that life on Earth had grown ever more complex, culminating in humankind—was another. The idea that humans had appeared recently—after Earth's topography, climate, and fauna had achieved their present form—was a third. These ideas interlocked to form a coherent, generalized picture of the past. They provided a context, shared by most educated Victorians, within which the findings of geology, paleontology, and anthropology could be discussed and given meaning.

This common context began to lose its explanatory power after 1859. The advent of Darwin's theory of evolution and the emergence of a new scientific consensus on human antiquity eroded the intellectual foundations of the common context. The rapid expansion and specialization of science narrowed the range scientific issues that scientists and laypeople could discuss on equal terms. Scientific and religious professionals alike called for science and religion to have separate but equal status, diminishing "official" support for natural theology.

During the 1860s, however, a new basis for understanding the past began to form within the scientific community. The new set of ideas and assumptions was more limited in scope and less universal in appeal than the older one. It attracted many scientists but comparatively fewer members of the clergy and the lay public. Within these narrower limits, however, it served many of the same functions as its predecessor. The new set of ideas provided a common foundation on which scientists from different disciplines could erect separate but linked explanations of the past. It served, too, as common intellectual ground on which career scientists and dedicated amateurs could meet and discuss their ideas about the past. The old common context encompassed the entire history of life; the new focused only on the latter stages of that history. The principal elements of the new view of the past included the coexistence of humans and extinct animals; the existence of a long prehistory in which human arts and institutions had been relatively primitive; and the idea that humankind had, in Europe, undergone a steady process of cultural (and perhaps physical) development.

The emergence of a new common context was part of the restructuring of Victorian science brought about by the establishment of human antiquity. The "great and sudden revolution" that began in Brixham Cave and ended in the pages of *Antiquity of Man* was only the beginning of the restructuring process. The establishment of human antiquity created the need for new relationships between disciplines while at the same time helping to erase the boundaries that had defined and separated existing disciplines. It was an end but equally, a beginning.

THE ESTABLISHMENT of human antiquity changed key elements in scientists' understanding of the (geologically) recent past. The most important of these changes was the erasure of a long-standing distinction between the modern world occupied by humans and the former worlds populated by extinct animals. Scientists had long seen the boundary between the two as a major landmark in both the history of the Earth and the history of life. Geologically, it marked the point at which major changes in the contours of the Earth's surface had ceased. Paleontologically, it marked the extinction of the mammoth, cave bear, and wooly rhinoceros and the appearance, in their place, of Europe's present-day fauna. The line between former and modern worlds formed the de facto boundary between the Post-Pliocene and Recent epochs. It also formed, less tangibly but no less significantly, the boundary between a familiar world and an alien one populated by monstrous creatures.

Acceptance of the idea that humans had lived in Post-Pliocene Europe entailed a different understanding of the relationship between *then* and *now*. It showed that the Europe known to the earliest humans was very different than that familiar to the Victorians: that it was a colder, harsher, and altogether "wilder" place. It also showed, however, that the world of the Paleolithic was continuous with that of the present day. No flood, upheaval, or great extinction separated the toolmakers of Abbeville from the savants of the Paris Academy of Sciences—only time.

Victorian scientists continued, long after 1863, to regard the appearance of humans as an important event in Earth history. They no longer saw it, however, as concurrent with the emergence of a world that was recognizably modern in topography, climate, flora, and fauna. The establishment of human antiquity permanently decoupled the two events. The existence of human beings, once a criterion for distinguishing between modern and former worlds, had become an element of continuity between them. That continuity, once established, became the basis of attempts to locate the earliest humans even further in the past. The presence of humans in the unfamiliar landscape of Europe's postglacial period made their presence in—or even before—the glacial period more plausible. Belief in Post-Pliocene Man did not necessarily entail belief in Pliocene Man, but did eliminate a wide range of a priori objections to the idea.

The emergence of a new consensus on human antiquity in 1858–63 changed the nature of the human antiquity question. It answered a sixty-year-old question, "Did humans coexist with extinct animals?" and opened a new one, "During what geological epoch did the first humans appear?" At the same time, it changed the context within which scientists investigated the age of the human race.

Two characteristics defined British scientists' work on human antiquity between 1820 and 1860. First, the work took place almost exclusively within the geological community. New evidence for human antiquity was examined by geologists, judged according to geological standards, and evaluated within the context of ongoing stratigraphic and paleontological research. Anthropologists and linguists sometimes put forward arguments of their own, but geologists controlled and evaluated the "hard evidence" of bones and tools.[2] Second, human antiquity research remained intermittent rather than continuous. Individual geologists' judgments on newly discovered evidence did not add up to an orga-

2. For one such argument, see D. K. Grayson, *The Establishment of Human Antiquity* (1983), chap. 7.

nized study of the problem. In the months or years between discoveries, interest in human antiquity languished.

Neither situation survived the events of 1858–63. The realization that humans had existed in Europe since the end of the
glacial period—and perhaps even longer—forced a revision of
ideas about the past. Once seen as a matter of centuries, European prehistory grew to encompass thousands upon thousands
of years. The expansion of prehistory posed major challenges for
Victorian scientists. It created a vast span of human history about
which little was known and which no existing discipline could
investigate in detail. The resulting intellectual vacuum, though it
made prehistoric archaeology a distinct field of inquiry for the
first time, was not filled by any single discipline. European prehistory became, instead, the subject of an interdisciplinary research tradition involving geology and anthropology as well as
archaeology.

Human antiquity research was enlarged, and its continuity
ensured, by its integration into the larger field of prehistoric studies. Beginning in the early 1860s, the age of the earliest humans
bore directly not only on geological questions but also on archaeological and cultural anthropological ones. The discovery, dating,
and explanation of early human remains became, by the mid-
1860s, a full-fledged research tradition. Interest in the subject no
longer waxed and waned as new evidence came to light but,
within a small but growing community of experts, remained constant.

The age of the human race remained a geological problem
after 1863, but it ceased to be solely a geological problem. The
difficult job of distinguishing between crude artifacts and naturally broken stones fell more and more frequently to archaeologists. The controversy over "eoliths" that dominated human
antiquity research in the 1880s and 1890s was, to a great degree, a dispute between geologists and archaeologists over the
proper interpretation of roughly broken flints. The task of identifying fossilized human bones was, in the same way, gradually
handed over to physical anthropologists. Geologists continued
to pass judgment on the stratigraphic position—and thus the
age—of bones and tools, but the changing nature of human an-

tiquity research made them one group of specialists among many.

MODERN HUMAN antiquity research reflects its nineteenth-century origins, but only imperfectly. The passage of more than a hundred years has changed methods, problems, and assumptions. Fossilized human bones, though never common, began to turn up in greater numbers during the late nineteenth century, and the physical anthropologist's role in human antiquity research grew accordingly. Beginning in the 1910s, scientists' conviction that human evolution was a branching process broadened the physical anthropologist's role still further. The search for "the earliest man" became a search for the earliest member of the evolutionary line leading to *Homo sapiens.* Determining whether newly discovered remains did, in fact, belong to an ancestor of modern humans required the specialized expertise of scientists now known as "paleoanthropologists." [3]

The cultural anthropologist's role in human antiquity research diminished as the physical anthropologist's grew, though for different reasons. The unilinear explanations of cultural development proposed by the classical evolutionists went into decline in the years around 1900. Younger anthropologists such as Franz Boas and Bronislaw Malinowski rejected the search for universal patterns of development and called instead for detailed studies of how particular primitive societies function.[4] The decline of classical evolutionism severed the once-intimate ties between anthropology and prehistoric archaeology. The anthropologists' shift in emphasis—from past to contemporary societies, and developmental to functional questions—eliminated the need for the detailed historical knowledge of particular cultures that archaeologists had once provided.

Geologists began, as early as the 1920s, to use the decay rates of radioactive isotopes as a chronometer for measuring the ages of rocks. Radiometric dating offered a solution to the long-insoluble

3. P. J. Bowler, *Theories of Human Evolution* (1986).
4. George Stocking has analyzed this process from several perspectives. See, e.g., "The Ethnographer's Magic," in *Observers Observed* (1983), and the epilogue to *Victorian Anthropology* (1987).

question of absolute age. Steadily refined and expanded (the development of the carbon 14 process in the 1940s allowed the dating of organic objects), radiometric dating has become an integral part of human antiquity research.[5] Answers to the question, "How old is the human race?" are now routinely expressed in numbers.

These changes are evident in Richard Leakey's research near Lake Turkana in Kenya, Donald Johanson's investigations in the Afar region of Ethiopia, and other recent work on early hominids. Organized and led by paleoanthropologists, recent expeditions have gone into the field with the intention of collecting human fossils and determining their absolute age through radiometric dating. The expeditions' results are awaited with keen interest by theorists, who regard them as essential in understanding human evolution. The methods, assumptions, and goals that Johanson and Leakey have brought into the field thus differ, in important ways, from those that Joseph Prestwich and James Geikie brought to their own research a century earlier. The disciplinary boundaries and interrelationships established in the 1860s are clearly visible, however, in modern investigations of human antiquity. Human antiquity research is still a cooperative enterprise, dependent on the efforts of geologists, paleontologists, archaeologists, anthropologists, and other specialists. The search for our earliest ancestors may have shifted from East Anglia to East Africa, but it still bears the stamp of its mid-Victorian origins.

5. For the history of radiometric dating, see C. C. Albritton, *The Abyss of Time* (1980), chap. 16. D. L. Eicher (*Geologic Time,* 2d ed. [1976], chap. 6); and G. Bibby (*The Testimony of the Spade* [1956], 214–17) offer helpful, nontechnical explanations of the relevant scientific principles. R. Lewin (*Bones of Contention* [1988], chaps. 9–10) discusses the significance of radiometric dating in modern paleoanthropology.

Contents of National Archaeological Journals, 1850–75

These three tables summarize the contents of papers published in Britain's leading archaeological journals between 1850 and 1875: *Archaeologia,* in table 1; *Archaeological Journal,* in table 2; and the *Journal of the British Archaeological Association,* in table 3. The papers are divided into five categories by the chronological period with which they deal (see note to table 1). I have not included abstracts, second-hand reports of papers presented at annual meetings, or brief items of "antiquarian intelligence." Multipart works that extended over several issues of a journal are counted as separate items.

TABLE 1 *Archaeologia* (Society of Antiquaries of London)

Year	A	B	C	D	E	Total
1849	0	5	2	18	1	26
1852	3	7	2	15	3	30
1853	2	4	7	21	1	35
1855	1	5	3	19	2	30
1857	2	1	4	23	0	30
1859-60	4	0	2	20	3	29
1863	1	3	1	19	3	27
1866	4	12	0	17	0	33
1867	1	7	3	12	2	25
1869	3	2	2	12	4	23
1871	1	4	0	11	2	18
1873	3	5	4	19	3	34
Total	25	55	30	206	24	340
Average per issue	2	5	2	17	2	28
All papers (%)	7.4	16.2	8.8	60.0	7.1	

NOTE.—*Archaeologia* was published irregularly; the issues listed here are those that cover, as nearly as possible, the twenty-five years between 1850 and 1875. A = Prehistoric (before 55 B.C.); B = Roman era (55 B.C.–410 A.D.); C = Saxon era (410 A.D.–900 A.D.); D = medieval or later (after 900 A.D.); E = non-European (any period).

TABLE 2 *Archaeological Journal* (Archaeological Institute)

Year	A	B	C	D	E	Total
1850	2	4	4	15	2	27
1851	1	6	2	17	1	27
1852	2	1	2	14	0	19
1853	2	2	0	20	0	24
1854	3	0	1	15	1	20
1855	0	3	0	17	0	20
1856	0	3	1	16	2	22
1857	0	3	1	17	0	21
1858	2	2	0	18	0	22
1859	1	4	1	13	1	20
1860	0	5	1	14	2	22
1861	3	0	0	16	2	21
1862	0	0	0	18	1	19
1863	1	1	1	16	1	20
1864	2	5	0	20	1	28
1865	2	3	1	9	1	16
1866	2	4	0	10	2	18
1867	5	4	1	13	0	23
1868	2	2	1	17	0	22
1869	2	2	0	16	0	20
1870	3	9	1	9	2	24
1871	3	5	2	14	0	24
1872	1	4	0	18	1	24
1873	2	9	1	13	1	26
1874	1	3	0	18	1	23
1875	0	2	1	16	0	19
Total	42	86	22	399	22	571
Average per issue	2	3	1	15	1	22
All papers (%)	7.3	15.1	3.8	69.8	3.9	

NOTE.—A = Prehistoric; B = Roman; C = Saxon era; D = medieval or later; E = non-European (see table 1 for inclusive dates of each period).

TABLE 3 *Journal of the British Archaeological Association*

year	A	B	C	D	E	Total
1850	0	7	1	14	0	22
1851	1	7	1	24	1	34
1852	1	6	0	28	2	37
1853	1	5	0	17	2	25
1854	2	4	0	16	0	22
1855a	2	2	5	10	0	19
1855b	1	2	3	13	1	20
1856	0	1	0	9	1	11
1857	1	4	2	6	2	15
1858	0	5	2	14	1	22
1859	—	—	—	—	—	—
1860	1	5	3	9	0	18
1861	1	4	1	15	0	21
1862	1	4	2	10	0	17
1863	0	4	0	9	0	13
1864	2	4	1	10	0	17
1865	1	2	2	18	1	24
1866	2	4	1	14	0	21
1867	1	3	4	10	1	19
1868	2	4	1	20	0	27
1869	0	2	1	16	1	20
1870	1	6	0	16	0	23
1871	1	3	0	22	1	27
1872	1	3	1	16	0	21
1873	1	8	2	11	0	22
1874	2	2	2	20	0	26
1875	1	3	3	23	0	30
Total	27	104	38	390	14	573
Average per issue	1	4	1.5	15	0.5	22
All Papers (%)	4.7	18.2	6.6	68.0	2.4	

NOTE.—A = Prehistoric; B = Roman; C = Saxon era; D = medieval or later; E = non-European (see table 1 for inclusive dates of each period).
    Two issues were published in 1855, none in 1859.

APPENDIX 2

Commentaries on Human Antiquity, 1859–75

There are 86 commentaries listed in table 4.

TABLE 4 Commentaries by Author

Author	Date	Title of Publication	Occu-pation	Type of Publication	Relative Age	Absolute Age
Adams, W. H. D.	1872	*Life in the ...*	Pub	Sci	3	5
Anderson, J.	1859	*BAAS Report*	Cler	Sci	2	—
Anonymous	1860	*British Quart. R.*	n/a	R/CB	1	—
Anonymous	1861	*Brit. & For. Evan. R.*	n/a	R/LA	3	2
Anonymous	1863	*Anthropological R.*	S/An	Sci	3	0
Anonymous	1863	*Athenaeum*	n/a	Gen	3	3
Anonymous	1863	*British Quart. R.*	n/a	R/CB	3	1
Anonymous	1863	*British Quart. R.*	n/a	R/CB	3	1
Anonymous	1863	*Christ. Rememb.*	n/a	R/HA	3	0
Anonymous	1863	*London Int. Obs.*	n/a	Gen	3	5
Anonymous	1863	*Saturday R.*	n/a	Gen	3	3
Anonymous	1863	*Westminster R.*	n/a	Gen	3	3
Anonymous	1864	*Ch. Family Mag.*	n/a	R/?	3	1
Anonymous	1865	*J. of Sacred Lit.*	n/a	R/HA	3	1
Anonymous	1868	*Eclectic R.*	n/a	R/CB	3	0
Anonymous	1869	*Brit. & For. Evan. R.*	n/a	R/LA	3	1
Anonymous	1874	*London Quart. R.*	n/a	R/Me	3	1
Ansted, D.	1860	*Geological Gossip*	S/Ge	Sci	3	0
Ansted, D.	1860	*National Review*	S/Ge	R/Un	3	5
Argyll, duke of	1862	*Pr. Royal Soc. Edin.*	S/Ot	Sci	3	5
Argyll, duke of	1869	*Primeval Man*	S/Ot	Sci	3	5
Babbage, C.	1859	*Proc. Royal Soc.*	S/Ma	Sci	2	—
Bate, S.	1867	*Popular Science R.*	S/Bi	Sci	3	0
Cooke, W.	1872	*Fallacies of the ...*	Cler	Rel	1	—
Crawfurd, J.	1863	*Anthropological R.*	S/An	Sci	3	3
Crawfurd, J.	1865	*Trans. Eth. Soc.*	S/An	Sci	3	3
D'Oyly, C. J.	1868	*Contemporary R.*	Cler	Gen	3	4
Darwin, C.	1871	*Descent of Man*	S/Bi	Sci	3	3
Dawkins, W. B.	1874	*British Quart. R.*	S/Ge	R/CB	3	5

TABLE 4   Continued

Author	Date	Title of Publication	Occu-pation	Type of Publication	Relative Age	Absolute Age
Dawson, J. W.	1860	*Archaia*	S/Ge	Sci	3?	1
Dawson, J. W.	1864	*Edin. New Phil. J.*	S/Ge	Sci	3	1
Dawson, J. W.	1873	*Story of the Earth*	S/Ge	Sci	3	1
Dawson, J. W.	1874	*J. Trans. Vict. Inst.*	S/Ge	Rel	3	1
Deane, G.	1874	*British Quart. R.*		R/CB	3	5?
Duncan, I.	1860	*Preadamite Man*	Pub	Rel	3	4
Essex Rector	1865	*Man's Age . . .*	Cler	Rel	3	2
Fergusson, J.	1870	*Quarterly R.*	S/Ar	Gen	3	0
Forbes, J.	1863	*Edinburgh R.*	S/Ge	Gen	3	2
Forbes, J.	1868	*Good Words*	S/Ge	R/LA	3	2
Haughton, S.	1861	*Dublin Univ. Mag.*	S/Ge	Gen	3	0
Henslow, J.	1859	*Athenaeum*	S/Bi	Gen	2	—
Henslow, J.	1860	*Athenaeum*	S/Bi	Gen	3	5
Henslow, J.	1860	*Athenaeum*	S/Bi	Gen	3	5
Henslow, J.	1860	*Athenaeum*	S/Bi	Gen	3	5
Henslow, J.	1861	*Edin. New Phil. J.*	S/Bi	Gen	3	5
Holland, H.	1864	*Edinburgh R.*	Phy	Gen	3	5
Hunt, J.	1866	*Pop. Mag. of Anth.*	S/An	Sci	3	3
Huxley, T. H.	1863	*Man's Place . . .*	S/Bi	Sci	3	3
Jukes, J.	1862	*Manual of Geology.*	S/Ge	Sci	3	0
Latham, R. G.	1860	*Westminster R.*	S/Ot	Gen	3	5
Lubbock, J.	1863	*Natural History R.*	S/Bi	Sci	3	3
Lysons, S.	1865	*Our Brit. Ancestors*	Cler	Sci	3	1
M'Causland, D.	1872	*Adam & Adamite*	Law	Rel	3	4
McLennan, J.	1869	*North British R.*	S/An	Gen	3	3
Moore, J. S.	1868	*Preglacial Man . . .*	Pub	Sci	3	4
Murchison, R.	1861	*BAAS Report*	S/Ge	Sci	3	0
Ogden, H.	1859	*Athenaeum*	Phy	Gen	1	—
Owen, R.	1860	*Paleontology*	S/Bi	Sci	3	0
Page, D.	1868	*Man: Where . . . ?*	S/Ge	Sci	3	3
Paley, F. A.	1864	*Home & Foreign R.*		R/RC	3	5
Parker, J.	1874	*J. Trans. Vict. Inst.*	Pub	Gen	3	2?
Pattison, S. R.	1868	*New Facts . . .*		Rel	3	1
Phillips, J.	1860	*Life on Earth*	S/Ge	Sci	3	0
Phillips, J.	1863	*BAAS Report*	S/Ge	Sci	3	2
Phillips, J.	1863	*Quarterly R.*	S/Ge	Gen	3	2
Pollock, W. F.	1863	*Fraser's Mag.*	Law	Gen	3	0
Portlock, J.	1859	*BAAS Report*	S/Ge	Sci	2	—
Rogers, H. D.	1860	*Blackwood's Mag.*	S/Ge	Gen	3	5
Smith, W.	1863	*Blackwood's Mag.*	Pub	Gen	3	2
Taylor, J.	1862	*Macmillan's Mag.*	S/Ge	Gen	2	—
Trevelyan, W. C.	1859	*Athenaeum*	S/Bi	Gen	2	—
Tylor, E.	1868	*Quarterly R.*	S/An	Gen	3	5
W. C.	1867	*Brit. & For. Evan. R.*	n/a	R/LA	3	1
Wallace, A.	1864	*BAAS Report*	S/Bi	Sci	3	3

TABLE 4   Continued

Author	Date	Title of Publication	Occupation	Type of Publication	Relative Age	Absolute Age
Wedgwood, J.	1863	*Macmillan's Mag.*	Pub	Gen	3	0
White, M.	1869	*Brit. & For. Evan. R.*		R/LA	3	0
Whitley, N.	1865	*Flint Implements*	Eng	Sci	1	—
Whitley, N.	1869	*Popular Science R.*	Eng	Sci	1	—
Whitley, N.	1874	*J. Trans. Vict. Inst.*	Eng	Rel	1	—
Whitley, N.	1874	*J. Trans. Vict. Inst.*	Eng	Rel	1	—
Wilkinson, G.	1860	*Athenaeum*	S/Ar	Gen	2	—
Wood, H.	1867	*Contemporary R.*	Cler	Gen	3	1?
Wright, T.	1859	*Athenaeum*	S/Ar	Gen	1	—
Wright, T.	1859	*Athenaeum*	S/Ar	Gen	1	—
Wrottesley, Lord	1860	*BAAS Report*	S/Ot	Sci	3	3
Young, J. R.	1865	*Mod. Scepticism . . .*	S/Ma	Rel	3	2

NOTE.—Consult bibliography for full citations. Abbreviations used:

Column 4: Occupation

Cler	Clergy	S/An	Anthropologist
Eng	Engineer	S/Ar	Archaeologist
Law	Lawyer	S/Bi	Biologist/naturalist
Phy	Physician	S/Ge	Geologist
Pub	Writer/publisher	S/Ma	Mathematician
n/a	Unknown	S/Ot	Other Scientist

Column 5: Type of publication

Gen	General Interest	R/CB	Congregationalist/Baptist
Sci	Scientific	R/M	Methodist
R/HA	Anglican (High Church)	R/RC	Roman Catholic
R/LA	Anglican (Low Church)	R/U	Unitarian
Rel	Religious (nondenominational)		

Column 6: Relative age

1   Argues that chipped flints are natural, not artificial
2   Argues that tool/bone associations are clearly accidental
3   Accepts coexistence of humans and extinct animals

Column 7: Absolute age

0   Does not discuss absolute age of human race
1   Traditionalist (6,000–8,000 years old)
2   Prestwichian (c. 20,000 years old)
3   Lyellian (100,000+ years old)
4   Accepts preadamite theory and defends it in detail
5   Argues that existing date is insufficient to settle the issue
—   Position on relative age makes discussion of absolute age moot

TABLE 5   Commentaries Grouped According to Theoretical Views

Author	Date	Title of Publication	Occu-pation	Type of Publication	Relative Age	Absolute Age
Wright, T.	1859	*Athenaeum*	S/AR	Gen	1	—
Wright, T.	1859	*Athenaeum*	S/Ar	Gen	1	—
Ogden, H.	1859	*Athenaeum*	Phy	Gen	1	—
Anonymous	1860	*British Quart. R.*	n/a	R/CB	1	—
Whitley, N.	1865	*Flint Implements*	Eng	Sci	1	—
Whitley, N.	1869	*Popular Science R.*	Eng	Sci	1	—
Cooke, W.	1872	*Fallacies of the . . .*	Cler	Rel	1	—
Whitley, N.	1874	*J. Trans. Vict. Inst.*	Eng	Rel	1	—
Whitley, N.	1874	*J. Trans. Vict. Inst.*	Eng	Rel	1	—
Anderson, J.	1859	*BAAS Report*	Cler	Sci	2	—
Babbage, C.	1859	*Proc. Royal Soc.*	S/Ma	Sci	2	—
Henslow, J.	1859	*Athenaeum*	S/Bi	Gen	2	—
Portlock, J.	1859	*BAAS Report*	S/Ge	Sci	2	—
Trevelyan, W. C.	1859	*Athenaeum*	S/Bi	Gen	2	—
Wilkinson, G.	1860	*Athenaeum*	S/Ar	Gen	2	—
Taylor, J.	1862	*Macmillan's Mag.*	S/Ge	Gen	2	—
Ansted, D.	1860	*Geological Gossip*	S/Ge	Sci	3	0
Owen, R.	1860	*Paleontology*	S/Bi	Sci	3	0
Phillips, J.	1860	*Life on Earth*	S/Ge	Sci	3	0
Haughton, S.	1861	*Dublin Univ. Mag.*	S/Ge	Gen	3	0
Murchison, R.	1861	*BAAS Report*	S/Ge	Sci	3	0
Jukes, J.	1862	*Manual of Geology.*	S/Ge	Sci	3	0
Anonymous	1863	*Anthropological R.*	S/An	Sci	3	0
Anonymous	1863	*Christ. Rememb.*	n/a	R/HA	3	0
Pollock, W. F.	1863	*Fraser's Mag.*	Law	Gen	3	0
Wedgwood, J.	1863	*Macmillan's Mag.*	Pub	Gen	3	0
Bate, S.	1867	*Popular Science R.*	S/Bi	Sci	3	0
Anonymous	1868	*Eclectic R.*	n/a	R/CB	3	0
White, M.	1869	*Brit. & For. Evan. R.*		R/LA	3	0
Fergusson, J.	1870	*Quarterly R.*	S/Ar	Gen	3	0
Dawson, J. W.	1860	*Archaia*	S/Ge	Sci	3?	1
Anonymous	1863	*British Quart. R.*	n/a	R/CB	3	1
Anonymous	1863	*British Quart. R.*	n/a	R/CB	3	1
Anonymous	1864	*Ch. Family Mag.*	n/a	R/?	3	1
Dawson, J. W.	1864	*Edin. New Phil. J.*	S/Ge	Sci	3	1
Anonymous	1865	*J. of Sacred Lit.*	n/a	R/HA	3	1
Lysons, S.	1865	*Our. Brit. Ancest.*	Cler	Sci	3	1
W. C.	1868	*Brit. & For. Evan. R.*	n/a	R/LA	3	1
Wood, H.	1867	*Contemporary R.*	Cler	Gen	3	1?
Pattison, S. R.	1868	*New Facts . . .*		Rel	3	1
Anonymous	1869	*Brit. & For. Evan. R.*	n/a	R/LA	3	1
Dawson, J. W.	1873	*Story of the Earth*	S/Ge	Sci	3	1

TABLE 5 Continued

Author	Date	Title of Publication	Occu-pation	Type of Publication	Relative Age	Absolute Age
Anonymous	1874	*London Quart. R.*	n/a	R/Me	3	1
Dawson, J. W.	1874	*J. Trans. Vict. Inst.*	S/Ge	Rel	3	1
Anonymous	1861	*Brit. & For. Evan. R.*	n/a	R/LA	3	2
Forbes, J.	1863	*Edinburgh R.*	S/Ge	Gen	3	2
Phillips, J.	1863	*BAAS Report*	S/Ge	Sci	3	2
Phillips, J.	1863	*Quarterly R.*	S/Ge	Gen	3	2
Smith, W.	1863	*Blackwood's Mag.*	Pub	Gen	3	2
Forbes, J.	1864	*Good Words*	S/Ge	R/LA	3	2
Essex Rector	1865	*Man's Age . . .*	Cler	Rel	3	2
Young, J. R.	1865	*Mod. Scepticism . . .*	S/Ma	Rel	3	2
Parker, J.	1874	*J. Trans. Vict. Inst.*	Pub	Gen	3	2?
Wrottesley, Lord	1860	*BAAS Report*	S/Ot	Sci	3	3
Anonymous	1863	*Athenaeum*	n/a	Gen	3	3
Anonymous	1863	*Saturday R.*	n/a	Gen	3	3
Anonymous	1863	*Westminster R.*	n/a	Gen	3	3
Crawfurd, J.	1863	*Anthropological R.*	S/An	Sci	3	3
Huxley, T. H.	1863	*Man's Place . . .*	S/Bi	Sci	3	3
Lubbock, J.	1863	*Natural History R.*	S/Bi	Sci	3	3
Wallace, A.	1864	*BAAS Report*	S/Bi	Sci	3	3
Crawfurd, J.	1865	*Trans. Eth. Soc.*	S/An	Sci	3	3
Hunt, J.	1866	*Pop. Mag. Anth.*	S/An	Sci	3	3
Page, D.	1868	*Man: Where . . .?*	S/Ge	Sci	3	3
McLennan, J.	1869	*North British R.*	S/An	Gen	3	3
Darwin, C.	1871	*Descent of Man*	S/Bi	Sci	3	3
Duncan, I.	1860	*Preadamite Man*	Pub	Rel	3	4
D'Oyly, C. J.	1868	*Contemporary R.*	Cler	Gen	3	4
Moore, J. S.	1868	*Preglacial Man . . .*	Pub	Sci	3	4
M'Causland, D.	1872	*Adam & Adamite*	Law	Rel	3	4
Ansted, D.	1860	*National Review*	S/Ge	R/Un	3	5
Henslow, J.	1860	*Athenaeum*	S/Bi	Gen	3	5
Henslow, J.	1860	*Athenaeum*	S/Bi	Gen	3	5
Henslow, J.	1860	*Athenaeum*	S/Bi	Gen	3	5
Latham, R. G.	1860	*Westminster R.*	S/Ot	Gen	3	5
Rogers, H. D.	1860	*Blackwood's Mag.*	S/Ge	Gen	3	5
Henslow, John	1861	*Edin. New Phil. J.*	S/Bi	Gen	3	5
Argyll, Duke of	1862	*Pr. Royal Soc. Edin.*	S/Ot	Sci	3	5
Anonymous	1863	*London Int. Obs.*	n/a	Gen	3	5
Holland, H.	1864	*Edinburgh R.*	Phy	Gen	3	5
Paley, F. A.	1864	*Home & Foreign R.*	n/a	R/RC	3	5
Tylor, E.	1868	*Quarterly R.*	S/An	Gen	3	5
Argyll, duke of	1869	*Primeval Man*	S/Ot	Sci	3	5

TABLE 5 Continued

Author	Date	Title of Publication	Occu-pation	Type of Publication	Relative Age	Absolute Age
Adams, W. H. D.	1872	*Life in the . . .*	Pub	Sci	3	5
Dawkins, W. B.	1874	*British Quart. R.*	S/Ge	R/CB	3	5
Deane, G.	1874	*Brisith Quart. R.*		R/CB	3	5?

NOTE.—For abbreviations used, see note to table 4.

BIBLIOGRAPHY

Manuscript Sources

I have not listed unpublished letters or field notebooks individually in the bibliography. Full citations for individual unpublished items are given in the notes, which the following material is designed to supplement.

Hugh Falconer. The Falconer Museum in Forres, Scotland, holds a substantial collection of manuscripts, including many letters to and from Joseph Prestwich. The letters have been cataloged and indexed by Patrick J. Boylan, whose catalog, "The Falconer Papers, Forres" (Leicester: Leicestershire Museums, Art Galleries and Records Service, 1977), is useful but unfortunately out of print. Items are cited as "Falconer Papers," followed by Boylan's index number.

Charles Lyell. The Edinburgh University Library holds a large collection of letters, filed alphabetically by author, with each author's letters arranged chronologically. The American Philosophical Society, in Philadelphia, holds a smaller collection of Lyell letters, individually numbered, including several on human antiquity. These are cited as "Lyell Papers, EUL" and "Lyell Papers, APS," respectively.

William Pengelly. The Torquay Natural Museum in Torquay, England, holds thirteen paperbound notebooks containing committee reports, journal entries, and copies of letters concerning the Brixham Cave excavation. The collection also includes Pengelly's field notes from the Kent's Cavern excavation of 1865–80 and an unindexed album of scientific correspondence. Items are cited as "Pengelly Papers."

Other Scientists. I have been unable to locate John Evans's papers, or the bulk of Joseph Prestwich's. Prestwich's field notebooks, in the library of the Geological Society of London, make no references to the human antiquity problem. John Lubbock's papers and diaries are housed at the

British Library but have little to say about human antiquity or its impact on archaeology. The Royal Society's library includes scattered items such as referees' reports on Prestwich's 1859 and 1862 papers on the Somme Valley, and letters concerning the final disposition of the specimens from Brixham Cave. The American Philosophical Society's "Papers of Scientists" and "Miscellaneous Manuscripts" collections include individual letters on human antiquity from John Evans, John Henslow, and T. H. Huxley.

Published Sources

Adams, W. H. Davenport. *Life in the Primeval World.* London: Nelson, 1872.

Agassiz, Elizabeth C., ed. *Louis Agassiz: His Life and Correspondence.* 2 vols. London: Macmillan, 1885.

Albritton, Claude C., Jr. *The Abyss of Time: Changing Conceptions of the Earth's Antiquity since the Sixteenth Century.* San Francisco: Freeman Cooper, 1980.

Allen, David E. "The Lost Limb: Geology and Natural History." In *Images of the Earth,* 199–212. See Jordanova and Porter, 1979.

———. *The Naturalist in Britain: A Social History.* Harmondsworth: Penguin, 1976.

Anderson, John. *The Course of Creation.* 1850. Cincinnati: Moore, Anderson, Wilstach, & Keys, 1853.

———. "Human Remains in Superficial Drift," *Annual Report of the British Association for the Advancement of Science* 29 (1859). Contents summarized in *Athenaeum,* October 1, 1859.

"The Antiquity of Man," *British and Foreign Evangelical Review* 16 (1867): 383–400.

"The Antiquity of Man." *British and Foreign Evangelical Review* 18 (1869): 288–315.

"The Antiquity of Man." *British Quarterly Review* 37 (1863): 410–41.

"The Antiquity of Man." *London Intellectual Observer.* Reprinted in *Eclectic Magazine* 59 (1863): 63–67.

"The Antiquity of Man." *London Quarterly Review* 42 (1874): 28–48.

"The Antiquity of Man." *Westminster Review* 79 (1863): 272–91.

Ansted, David T. *Geological Gossip.* London: Routledge, 1860.

[Ansted, David T.] "The Testimony of Geology to the Age of the Human Race." *National Review* 10 (1860): 279–312.

Argyll, duke of. "Opening Address." *Proceedings of the Royal Society of Edinburgh* 4 (1862): 362–70.

———. *Primeval Man.* New York: George Routledge & Sons, 1869.

Babbage, Charles. "Observations on the Discovery in Various Localities of the Remains of Human Art Mixed with the Bones of Extinct Races of Animals." *Proceedings of the Royal Society* 10 (1858): 59–72.

Babington, Churchill. *Introductory Lecture on the Study of Archaeology Delivered before the University of Cambridge.* Cambridge: Dayton & Bell, 1865.

Barber, Lynn. *The Heyday of Natural History.* New York: Doubleday, 1980.

Bartholomew, Michael J. "Lyell and Evolution: An Account of Lyell's Response to the Prospect of an Evolutionary Ancestry for Man," *British Journal for the History of Science* 6 (1973): 261–303.

Bate, Spence. "An Attempt to Approximate the Date of the Flint Flakes of Devon and Cornwall." *Popular Science Review.* Reprinted in *Eclectic Magazine* 63 (1867): 759–68.

Berry, W. B. N. *The Growth of a Prehistoric Time Scale.* 2d ed. Palo Alto, CA: Blackwell, 1987.

Bibby, Geoffrey. *The Testimony of the Spade.* London: Collins, 1956.

Bonney, T. G. "The Scientific Work of William Pengelly, FRS." In *A Memoir of William Pengelly,* edited by Hester Pengelly, 301–14. London: Murray, 1897.

Borlase, William Copeland. "Account of the Exploration of Tumuli at Tevalga, in the Parish of St. Coulomb Minor, Cornwall." *Archaeologia* 24 (1873): 422–27.

———. *Naenia Cornubiae.* London, 1872.

Bowden, Mark. *Pitt Rivers.* Cambridge: Cambridge University Press, 1991.

Bowler, Peter J. *Fossils and Progress: Paleontology and the Idea of Progressive Evolution in the Nineteenth Century.* New York: Science History, 1976.

———. *The Invention of Progress: The Victorians and the Past.* Oxford: Blackwell, 1990.

———. *Theories of Human Evolution: A Century of Debate, 1844–1944.* Baltimore: Johns Hopkins University Press, 1986.

Boylan, Patrick J. "The Controversy of the Moulin-Quignon Jaw: The Role of Hugh Falconer." In *Images of the Earth,* 171–99. See Jordanova and Porter, 1979.

Brice, William R. "Bishop Ussher, John Lightfoot and the Age of Creation." *Journal of Geological Education* 30 (1982): 18–24.

Brockie, William. "More Flint Implements." *Athenaeum,* December 17, 1859: 815.

Brooke, John Hedley. "The Natural Theology of Geologists: Some Theological Strata." In *Images of the Earth,* 39–62. See Jordanova and Porter, 1979.

———. *Science and Religion: Some Historical Perspectives.* Cambridge: Cambridge University Press, 1991.

Bruce, John C. "The Practical Advantages Accruing from the Study of Archaeology." *Archaeological Journal* 14 (1857): 1–7.

Buckland, William. *Geology and Mineralogy Considered with Reference to Natural Theology.* 2 vols. 1836. First American edition, Philadelphia: Carey, Lea, & Blanchard, 1837.

———. *Geology and Mineralogy Considered with Reference to Natural Theology.* 2d ed. 2 vols. London: Routledge, 1858.

Burchfield, Joe D. *Lord Kelvin and the Age of the Earth.* 1975. Chicago: University of Chicago Press, 1990.

Burrow, J. W. *Evolution and Society.* Cambridge: Cambridge University Press, 1966.

Bynum, W. F. "Charles Lyell's *Antiquity of Man* and Its Critics." *Journal of the History of Biology* 17 (1984): 153–87.

Carnarvon, earl of. "The Archaeology of Berkshire." *Journal of the British Archaeological Association* 15 (1860): 1–25.

Chadwick, Owen. *The Victorian Church.* 2 vols. New York: Oxford University Press, 1967–70.

Chandler, Alice. *A Dream of Order: The Medieval Ideal in Nineteenth-Century English Literature.* Lincoln: University of Nebraska Press, 1970.

Chapman, William Ryan. "Arranging Ethnology: A. H. L. F. Pitt-Rivers and the Typological Tradition." In *Objects and Others: Essays on Museums and Material Culture,* ed. George W. Stocking, Jr., 15–48. Madison: University of Wisconsin Press, 1985.

———. "The Organizational Context in the History of Archaeology: Pitt Rivers and Other British Archaeologists in the 1860s." *Antiquaries' Journal* 69 (1989): 23–42.

Chippindale, Christopher. "'Social Archaeology' in the Nineteenth Century: Is It Right to Look for Modern Ideas in Old Places?" In *Tracing Archaeology's Past,* ed. Andrew L. Christenson, 21–33. Carbondale: Southern Illinois University Press, 1989.

Clark, L. K. *Pioneers of Prehistory in England.* London: Sheed & Ward, 1961.

Cohen, Claudine, and Jean-Jacques Hublin. *Boucher De Perthes (1788–1868): Les origines romantiques de la préhistoire.* Paris: Belin, 1989.

Collins, H. M. "The Place of the 'Core Set' in Modern Science," *History of Science* 19 (1981): 6–19.

Conyngham, Albert. "Presidential Address," *Journal of the British Archaeological Association* 5 (1849): 285–88.

Cooke, William. *The Fallacies of the Alleged Antiquity of Man Proved.* London: Hamilton, Adams, 1872.

Crawfurd, John. "Notes on Sir Charles Lyell's Antiquity of Man." *Anthropological Review* 1 (1863): 172–76.

———. "On Lyell's 'Antiquity of Man' and on Prof. Huxley's 'Evidence as to Man's Place in Nature.'" *Transactions of the Ethnological Society of London,* n.s., 3 (1865): 58–70.

Croker, Dillon. "On the Advantages of the Study of Archaeology." *Journal of the British Archaeological Association* 5 (1849): 288–89.

Crowther, M. A. *Church Embattled: Religious Controversy in Mid-Victorian England.* London: David & Charles, 1970.

Culler, A. Dwight. *The Victorian Mirror of History.* New Haven, CT: Yale University Press, 1985.

Cuming, H. Syer. "On an Ancient British Snow Knife." *Journal of the British Archaeological Association* 24 (1868): 125–28.

Cunnington, Robert H. *From Antiquary to Archaeologist.* Aylesbury: Shire Publications, 1975.

D'Oyly, C. J. "Man in Creation." *Contemporary Review* 8 (1868): 550–68.

Daniel, Glyn. *The Idea of Prehistory.* London: Watts, 1962.

———. *150 Years of Archaeology.* London: Duckworth, 1975.

———, ed. *Towards a History of Archaeology.* London: Thames & Hudson, 1981.

Darwin, Charles. *The Autobiography of Charles Darwin.* Edited by Nora Barlow. London: Collins, 1958.

———. *The Descent of Man.* 1871. Reprint. Princeton, NJ: Princeton University Press, 1981.

Darwin, Francis. *The Life and Letters of Charles Darwin.* 3 vols. London: Murray, 1887.

Davies, Gordon L. *The Earth in Decay: A History of British Geomorphology, 1578–1878.* London: Macdonald, 1969.

Dawkins, William Boyd. *Cave Hunting.* London: Macmillan, 1874.

———. *Early Man in Britain.* London: Macmillan, 1880.

———. Preface. In *The Life and Letters of William Buckland.* 2 vols. Edited by Elizabeth O. Gordon. London: Murray, 1894.

[Dawkins, William Boyd.] "The Antiquity of Man." *British Quarterly Review* 58 (1874): 342–67.

Dawson, J. W. *Archaia.* 2d ed. London: Sampson Low, 1860.

———. *Fossil Men and Their Modern Representatives.* 3d ed. London: Hodder & Stoughton, 1887.

———. "Primitive Man and Revelation." *Journal of the Transactions of the Victoria Institute* 8 (1874–75): 59–63.

———. *The Story of the Earth and Man.* London: Hodder & Stoughton, 1873.

[Dawson, J. W.] "On the Antiquity of Man: A Review of 'Lyell' and 'Wilson.'" *Edinburgh New Philosophical Journal,* n.s. 19 (1864): 40–64.

De la Beche, Henry T. *A Geological Manual.* London, 1832.

———. *The Geological Observer.* 2nd ed. London: Longman, 1853.

De Laet, Siegfried J. "Phillipe-Charles Schmerling (1791–1836)." In *Towards a History of Archaeology,* 112–19. See Daniel, 1981.

Deane, George. "Modern Scientific Inquiry and Religious Thought." *British Quarterly Review* 59 (1874): 34–54.

Deane, Phyllis. *The First Industrial Revolution.* 2d ed. Cambridge: Cambridge University Press, 1979.

Desmond, Adrian. *Archetypes and Ancestors: Paleontology in Victorian London, 1850–75.* Chicago: University of Chicago Press, 1982.

Di Gregorio, Mario. *T. H. Huxley's Place in Natural Science.* New Haven, CT: Yale University Press, 1984.

"Dillettante Science . . . Made Easy." *Churchman's Family Magazine* 4 (1864): 472–78.

Duncan, Isabella. *Pre-Adamite Man.* 2d ed. London: Saunders, Otley, 1860.

Eicher, Don L. *Geologic Time.* 2d ed. Englewood Cliffs, NJ: Prentice Hall, 1976.

Ellegard, Alvar. *Darwin and the General Reader.* 1957. Reprint. Chicago: University of Chicago Press, 1989.

Essex Rector. *Man's Age in the World: According to Holy Scripture and Science.* London: Lovell-Reeve, 1865.

Evans, Joan. *A History of the Society of Antiquaries of London.* London: Society of Antiquaries, 1956.

———. "The Royal Archaeological Institute: A Retrospect." *Archaeological Journal* 56 (1949): 1–11.

————. *Time and Chance: The Story of Arthur Evans and His Forbears.* London: Longmans, 1943.

Evans, John. "The Abbeville Human Jaw." *Athenaeum,* June 6, 1863: 747–78.

————. "Account of Some Further Discoveries of Flint Implements in the Drift on the Continent and in England." *Archaeologia* 39 (1862): 57–84.

————. *The Ancient Stone Implements, Weapons, and Ornaments of Great Britain.* New York: Appleton, 1872.

————. "Flint Implements in the Drift." *Athenaeum,* June 25, 1859: 841.

————. "The Human Remains at Abbeville." *Athenaeum,* July 4, 1863: 19–20.

Evans, John. "On the Occurrence of Flint Implements in Undisturbed Beds of Gravel, Sand, and Clay." *Archaeologia* 38 (1860): 280–307.

Falconer, Hugh. "Falconer on the Reputed Fossil Man of Abbeville." *Anthropological Review* 1 (1863): 177–79.

————. "Primeval Man and His Contemporaries." In *Paleontological Memoirs and Notes of Hugh Falconer,* 2:571–600. See Murchison, 1868.

Falconer, Hugh, G. Busk, and W. B. Carpenter. "An Account of the Proceedings of the Late Conference Held in France to Inquire into the Circumstances Attending to the Asserted Discovery of a Human Jaw in the Gravel at Moulin-Quignon, near Abbeville." *Natural History Review* 3 (1863): 432–63.

[Fergusson, James.] "Non-Historic Times." *Quarterly Review* 128 (1870): 432–73.

Fichman, Martin. *Alfred Russel Wallace.* Boston: Twayne, 1981.

Forbes, James D. "On the Antiquity of Man." *Good Words* 5 (1864): 253–58, 432–40.

[————.] "Lyell on the Antiquity of Man." *Edinburgh Review* 118 (1863): 255–302.

Frere, John. "Account of Flint Weapons Discovered at Hoxne in Suffolk." *Archaeologia* 13 (1800): 204–5.

Geikie, Archibald. "Anniversary Address." *Proceedings of the Geological Society of London* 47 (1891): 48–162.

Geikie, James. *The Great Ice Age and Its Relation to the Antiquity of Man.* 1874. New York: Appleton, 1879.

Gillespie, Neil R. "The Duke of Argyll, Evolutionary Anthropology, and the Art of Scientific Controversy" *Isis* 68 (1977): 40–54.

Gillispie, Charles C. *Genesis and Geology*, 1951. New York: Harper Torch-books, 1959.

Girouard, Mark. *Life in the English Country House*. New Haven, CT: Yale University Press, 1978.

———. *Return to Camelot*. New Haven, CT: Yale University Press, 1981.

Glick, Thomas F., ed. *The Comparative Reception of Darwinism*. 1974. Reprint. Chicago: University of Chicago Press, 1988.

Gorilla [Phillip Egerton]. "Monkeyana." *Punch*, May 18, 1861: 206.

Gräslund, Bo. "Thomsen's Three-Age System." In *Towards History of Archaeology*. See Daniel, 1981.

Grayson, Donald K. *The Establishment of Human Antiquity*. New York: Academic Press, 1983.

———. "The First Three Editions of Charles Lyell's *Geological Evidences of the Antiquity of Man*." *Archives of Natural History* 13 (1985): 105–21.

———. "Nineteenth Century Explanations of Pleistocene Extinctions: A Review and Analysis." In *Quaternary Extinctions*, edited by P. S. Martin and R. G. Klein. Tucson: University of Arizona Press, 1983.

Greenwell, William. "Notices of the Examination of Ancient Grave-Hills in the North Riding of Yorkshire." *Archaeological Journal* 22 (1865): 97–117, 241–65.

Gregory, Frederick. "The Impact of Darwinian Evolution on Protestant Theology in the Late Nineteenth Century." In *God and Nature*, 369–90. See Lindberg and Numbers, 1986.

Gruber, Jacob W. "Brixham Cave and the Antiquity of Man." In *Context and Meaning in Cultural Anthropology*, edited by Melford E. Spiro, 373–402. New York: Free Press, 1965.

Haughton, Samuel. "The History of the Earth and Its Inhabitants." *Dublin University Magazine* 58 (1861): 105–113.

Henslow, John S. "Celts in the Drift." *Athenaeum*, November 19, 1859: 668.

———. "Flint Weapons in the Drift." *Athenaeum*, February 11, 1860: 206–7.

———. "Flints in the Drift." *Athenaeum*, October 20, 1860: 516.

———. Report of Public Lecture at Ipswich. *Edinburgh New Philosophical Journal* 14 (1861): 171.

———. "Works of Art in the Drift." *Athenaeum*, December 24, 1859: 853.

[Holland, Henry.] "Man and Nature." *Edinburgh Review* 120 (1864): 240–59 in American edition; 464–500 in British edition.

Hopkins, William. "Address of the President." *Quarterly Journal of the Geological Society of London* 8 (1852): xxiv–lxxx.

Horner, Leonard. "Address of the President." *Quarterly Journal of the Geological Society of London* 3 (1847): xxii–xc.

Hoskyns, Chandos Wren. "Inaugural Address at the Hereford Congress." *Journal of the British Archaeological Association* 27 (1871): 1–45.

Houghton, Lord. "Inaugural Address Delivered at the Opening of the Congress Held at Leeds." *Journal of the British Archaeological Association* 20 (1864): 1–15.

Hudson, Kenneth. *A Social History of Archaeology.* London: Macmillan, 1981.

Hull, David L., ed. *Darwin and His Critics.* Chicago: University of Chicago Press, 1973.

Hunt, James. "The Principles of Archaic Anthropology." *Popular Magazine of Anthropology* 3. Appended to *Anthropological Review* 4 (1866): 89–93.

Hutchinson, Horace G. *The Life and Letters of Sir John Lubbock.* 2 vols. London: Macmillan, 1914.

Huxley, Leonard. *The Life and Letters of T. H. Huxley.* 2 vols. New York: Appleton, 1901.

Huxley, T. H. "On The Relation of Man to the Lower Animals." In *Man's Place in Nature and Other Anthropological Essays,* 157–209. 1863. New York: Appleton, 1899.

———. "On Some Fossil Remains of Man." In *Man's Place in Nature and Other Anthropological Essays,* 77–156.

Jenkyns, Richard. *The Victorians and Ancient Greece.* Oxford: Blackwell, 1980.

Jenyns, Leonard. *Memoir of the Reverend John Stevens Henslow.* London: Van Voorst, 1862.

Jessup, Ronald. *Man of Many Talents.* Chichester: Phillimore, 1975.

Jewitt, Llewellyn, *Grave Mounds and Their Contents: A Manual of Archaeology.* London: Groombridge, 1870.

Jordanova, L. J., and Roy Porter, eds., *Images of the Earth.* Chalfont St. Giles: British Society for the History of Science, 1979.

Jukes, Joseph Beete. *Student's Manual of Geology.* 2d ed. Edinburgh: Black, 1862.

Keith, Arthur. "Anthropology." In *The Life-Work of Lord Avebury (Sir John Lubbock),* edited by Ursula Grant Duff, 67–104. London: Watts, 1924.

Kemble, J. M. "On the Utility of Antiquarian Collections in Relation to the Pre-historic Annals of the Different Countries of Europe," *Proceedings of the Royal Irish Academy* 6 (1857): 462–80.

[Kemble, J. M.] "Introduction." *Archaeological Journal* 6 (1849): 1–3.

Kennard, A. S. "The Early Digs in Kent's Hole, Torquay, and Mrs. Cazalet." *Proceedings of the Geologists' Association* 56 (1945): 156–213.

Klindt-Jensen, Ole. *A History of Scandinavian Archaeology.* London: Thames & Hudson, 1975.

Landau, Misia. *Narratives of Human Evolution.* New Haven, CT: Yale University Press, 1991.

[Latham, R. G.] "Antiquity of the Human Race." *Westminster Review* (American ed.) 74 (1860): 219–33.

Laudan, Rachel. *From Mineralogy to Geology: The Origins of a Science, 1660–1830.* Chicago: University of Chicago Press, 1987.

———. "Ideas and Organizations in British Geology: A Case Study in Institutional History." *Isis* 68 (1977): 527–38.

Levine, Philippa J. A. *The Amateur and the Professional.* Cambridge: Cambridge University Press, 1986.

Lewin, Roger. *Bones of Contention: Controversies in the Search for Human Origins.* New York: Touchstone–Simon & Schuster, 1988.

Lindberg, David C., and Ronald L. Numbers, eds. *God and Nature: Historical Essays on the Encounter between Christianity and Science.* Berkeley and Los Angeles: University of California Press, 1986.

Livingstone, David N. "The Preadamite Theory and the Marriage of Science and Religion." *Transactions of the American Philosophical Society* 82, no. 3, 1992.

Lubbock, John. "Address Delivered to the Section of 'Primeval Antiquities' at the London Meeting of the Archaeological Institute, July, 1866." *Archaeological Journal* 23 (1867): 190–209.

———. *Pre-historic Times.* London: Williams & Norgate, 1865.

[———.] Review of *Antiquity of Man,* by Charles Lyell. *Natural History Review* 3 (1863): 211–19.

Lyell, Charles. *Geological Evidences of the Antiquity of Man.* London: Murray, 1863.

———. *A Manual of Elementary Geology.* 5th ed. London: Murray, 1855.

———. "On the Occurrence of Works of Human Art in Post-Pliocene Deposits." *Annual Report of the British Association for the Advancement of Science* 29 (1859): 93–95.

———. "Presidential Address." *Annual Report of the British Association for the Advancement of Science* 34 (1864): lx–lxxv.

Lyell, Katherine M. *Life and Letters of Sir Charles Lyell.* 2 vols. London: Murray, 1881.

"Lyell on the Geological Evidence of the Antiquity of Man." *Anthropological Review* 1 (1863): 124–36.

Lyon, John. "Immediate Reactions to Darwin: The English Catholic Press' First Reviews of the *Origin of Species.*" *Church History* 41 (1972): 78–93.

———. "The Search for Fossil Man: Cinq personnages à la recherche du temps perdu." *Isis* 61 (1970): 68–84.

Lysons, Samuel. *Our British Ancestors: Who and What Were They?* Oxford: Henry & Parker, 1865.

Lytton, Lord. "Inaugural Address at the St. Albans Congress." *Journal of the British Archaeological Association* 26 (1870): 1–33.

M'Causland, Dominick. *Adam and the Adamite.* 3d ed. London: Richard Bentley, 1872.

MacEnery, John. *Cavern Researches.* Edited by Edward Vivian. London: Simpkin, Marshall, 1859.

McKinney, H. Lewis. *Wallace and Natural Selection.* New Haven, CT: Yale University Press, 1972.

[McLennan, J. F.] "The Early History of Man." *North British Review* 50 (1869): 272–90.

Mantell, Gideon. "Illustrations of the Connexion between Archaeology and Geology." *Edinburgh New Philosophical Journal* 50 (1851): 235–54.

———. *Medals of Creation.* 2d ed. 2 vols. 1854. Reprint. New York: Arno, 1980.

———. *The Wonders of Geology.* Edited by T. Rupert Jones. 2 vols. 7th ed. London: Bohn, 1857–58.

Marsden, Barry M. *The Early Barrow Diggers.* Aylesbury: Shire Publications, 1974.

———. *Pioneers of Prehistory.* Ormskirk: Hesketh, 1984.

Meltzer, David J. "The Antiquity of Man and the Development of American Archaeology," *Advances in Archaeological Method and Theory* 6 (1983): 1–51.

Merrill, Lynn. *The Romance of Victorian Natural History.* Oxford: Oxford University Press, 1989.

Michell, John. *Megalithomania: Artists, Antiquarians, and Archaeologists at the Old Stone Monuments.* London: Thames & Hudson, 1982.

"Modern Anthropology." *British Quarterly Review* 38 (1861) 466–98.

Moir, Esther. *The Discovery of Britain.* London: Routledge, 1964.

Moore, J. Scott. *Pre-Glacial Man and Geological Chronology.* Dublin: Hodges, Smith, & Foster, 1868.

Moore, James R. "Geologists and Interpreters of Genesis in the Nineteenth Century." In *God and Nature,* 322–50. See Lindberg and Numbers, 1986.

———. *The Post-Darwinian Controversies.* Cambridge: Cambridge University Press, 1982.

Morlot, Adolphe. "An Introductory Lecture to the Study of High Antiquity." *Smithsonian Annual Report for 1862* (Washington: U.S. Government Printing Office, 1863): 303–15.

Morrell, Jack. "Science and Government: John Phillips (1800–1874) and the Early Ordnance Survey of Great Britain." In *Science, Politics, and the Public Good,* edited by Nicolaas A. Rupke, 7–29. London: Macmillan, 1988.

Mortimer, John R. *Forty Years' Digging in the Grave Hills of Yorkshire.* London, 1905.

Mount Edgcumbe, Earl of. "Inaugural Address at the Cornwall Congress." *Journal of the British Archaeological Association* 33 (1877): 1–14.

Murchison, Charles. "Biographical Sketch." In *Paleontological Memoirs and Notes of Hugh Falconer,* 1: xxiii–liii.

———, ed. *Paleontological Memoirs and Notes of Hugh Falconer.* 2 vols. London: Hardwicke, 1868.

Murchison, Roderick I. "Thirty Years Retrospect of the Progress in Our Knowledge of the Geology of the Older Rocks: Presidential Address to Section C, BAAS, 1861." *American Journal of Science* 83 (1862): 1–21.

Newton, Charles. "On the Study of Archaeology." *Archaeological Journal* 8 (1850): 1–26.

North, F. J. "Paviland Cave, the 'Red Lady,' the Deluge, and William Buckland." *Annals of Science* 5 (1942), 91–128.

Northampton, marquis of. "Inaugural Address." *Archaeological Journal* 2 (1846): 302–3.

Northcote, Stafford H. "Inaugural Address at the Exeter Congress." *Journal of the British Archaeological Association* 18 (1862): 1–21.

Oakley, Kenneth P. "The Problem of Man's Antiquity." *Bulletin of the British Museum (Natural History), Geological Series* 9 (1964): 86–153.

Ogden, Henry M. "The Flint Find." *Athenaeum,* November 5, 1859: 666.

Oldfield, Edmund. "Introductory Address." *Archaeological Journal* 10 (1852): 1–6.

Oldroyd, David R. *The Highlands Controversy: Constructing Geological Knowledge through Fieldwork in Nineteenth-Century Britain.* Chicago: University of Chicago Press, 1990.

Owen, Richard. *Paleontology.* Edinburgh: Adam & Charles Black, 1860.

Page, David. *Man: Where, Whence, and Whither?* New York: Moorehead, Simpson, & Bond, 1868.

Parker, James. "The Flint Implements in the Valley of the Somme." *Journal of the Transactions of the Victoria Institute* 8 (1874–75): 51–58.

Pattison, S. R. *New Facts and Old Records: A Plea for Genesis.* London: Jackson, 1868(?).

Pengelly, Hester. *A Memoir of William Pengelly.* London: Murray, 1897.

Pengelly, William. *An Introductory Address Delivered before the Torquay Natural History Society at the Commencement of the Lecture Session, December 7, 1863.* London: Simpkin & Marshall, 1864.

———. *Kent's Cavern: Its Testimony to the Antiquity of Man.* Glasgow: Collins, 1876.

———. *Report on Two Lectures Delivered at Worcester and Malvern, Wednesday, January 26, and Friday, January 28, 1870, on the Ancient Cave Men of Devon.* Worcester: *Worcester Herald,* 1870.

Perkin, Harold. *The Origins of English Society.* London: Routledge, 1969.

Pettigrew, T. J. "On the Antiquities of the Isle of Wight." *Journal of the British Archaeological Association* 11 (1855): 177–93.

———. "On the Study of Archaeology and the Objects of the British Archaeological Association." *Journal of the British Archaeological Association* 6 (1850): 163–76.

Phillips, John. *Life on Earth.* London: Macmillan, 1860.

———. *Manual of Geology.* London: Richard Griffith, 1855.

———. "On the Deposit . . . at St. Acheul." *Annual Report of the British Association for the Advancement of Science* 33 (1863): 85–86.

———."Presidential Address to Section C." *Annual Report of the British Association for the Advancement of Science* 34 (1864): 45–49.

[Phillips, John.] Review of *Geological Evidences of the Antiquity of Man,* by Charles Lyell. *Quarterly Review* 114 (1863): 369–417.

Piggott, Stuart. *Ancient Britons and the Antiquarian Imagination*. New York: Thames & Hudson, 1989.

———. "The Origin of the English County Archaeological Societies." In *Ruins in a Landscape: Essays in Antiquarianism*, 171–96 Edinburgh: Edinburgh University Press, 1976.

———. "Prehistory and the Romantic Movement." *Antiquity* 11 (1937): 31–38.

———. *William Stukeley*. Rev. ed. New York: Thames & Hudson, 1985.

[Pollock, W. F.]. Review of *Geological Evidences of the Antiquity of Man*, by Charles Lyell. *Fraser's* 67 (1863): 463–75.

Poole, R. S., ed. *The Genesis of the Earth and of Man*. London: Williams & Norgate, 1860.

Popkin, Richard. "Pre-Adamism in Nineteenth Century American Thought: 'Speculative Biology' and Racism." *Philosophia* 8 (1978–79): 205–39.

Porter, Roy. "Charles Lyell: The Public and Private Faces of Science." *Janus* 69 (1982): 29–50.

———. "Creation and Credence: The Career of Theories of the Earth in Britain." In *Natural Order*, edited by Barry Barnes and Stephen Shapin. London: Sage, 1980.

———. "Gentlemen and Geology: The Emergence of a Scientific Career, 1660–1940." *Historical Journal* 21 (1978): 809–36.

———. *The Making of Geology: Earth Sciences in Britain, 1660–1815*. 1977. Cambridge: Cambridge University Press, 1980.

[Portlock, Joseph.] Commentary on Papers Read before Section C, BAAS. *Athenaeum*, October 8, 1859: 470.

"Prehistoric Man." *Christian Remembrancer* 46 (1863): 348–85.

Prestwich, Grace McCall. *Life and Letters of Sir Joseph Prestwich*. Edinburgh: Blackwood, 1899.

———. "Recollections of M. Boucher de Perthes." In *Essays Descriptive and Biographical*, 73–92. London: Blackwood, 1901.

Prestwich, Joseph. "Flint Implements in the Drift." *Athenaeum*, December 3, 1859: 740–41.

———. "Flint Implements in the Drift." *Athenaeum*, December 10, 1859: 775–76.

———. "On the Age, Formation, and Successive Drift Stages of the Valley of Darent; with Remarks on the Paleolithic Implements of the District, and on the Origin of Its Chalk Escarpment." *Quarterly Journal of the Geological Society of London* 47 (1891): 126–60.

————. "On the Occurrence of Flint-implements, associated with the Remains of Animals of Extinct Species in Beds of a late Geological Period, in France at Amiens and Abbeville, and in England at Hoxne." *Philosophical Transactions of the Royal Society of London* 150 (1860): 277–318.

————. "On the Primitive Characters of the Flint Implements of the Chalk Plateau of Kent, with Reference to the Question of Their Glacial or Pre-Glacial Age." *Journal of the Royal Anthropological Institute* 21 (1892): 246–62.

————. "On the Quaternary Flint Implements of Abbeville, Amiens, Hoxne &c., Their Geological Position and History." *Proceedings of the Royal Institution* 4 (1864): 213–22.

————. "The Past and Future of Geology." In *Smithsonian Annual Report for 1875*, 175–95. Washington: U.S. Government Printing Office, 1876.

————. "Theoretical Considerations on . . . the Drift Deposits Containing the Remains of Extinct Mammalia and Flint Implements." *Philosophical Transactions of the Royal Society of London* 154 (1864): 247–309.

Prestwich, Joseph, G. Busk, and J.[ohn] Evans. "Report on the Excavation of Brixham Cave." *Philosophical Transactions of the Royal Society of London* 163 (1873): 471–570.

"[Proceedings of] Archaeological and Ethnological Societies." *Geologist* 4 (1861), 153–55.

"Proceedings at Meetings of the Archaeological Institute." *Archaeological Journal* 17 (1860), 169–72.

Ramsay, Andrew C. "Works of Art in the Drift." *Athenaeum*, July 16, 1859: 83.

Reader, John. *Missing Links: The Hunt for Earliest Man.* 2d ed. Harmondsworth: Pelican, 1988.

"Recent Geological Speculations." *British and Foreign Evangelical Review* 10 (1861): 885–904.

Review of *The Genesis of the Earth and of Man,* edited by R. S. Poole. *Journal of Sacred Literature* 12 (1860): 123–33.

Review of *Geological Evidences of the Antiquity of Man,* by Charles Lyell. *Athenaeum*, February 14, 1863: 219–21.

Review of *Geological Evidences of the Antiquity of Man,* by Charles Lyell. *Saturday Review*, March 7, 1863: 311–12.

Review of *Geological Gossip,* by D. T. Ansted. *British Quarterly Review* 32 (1860): 258–9.

Review of *Geology and Mineralogy,* 2d ed., by William Buckland. *Dublin University Magazine* 54 (1859): 455–69.

Review of *Man's Age in the World,* by "An Essex Rector." *Journal of Sacred Literature* 8 (1866): 477–79.

Review of *A Student's Manual of Geology,* by J. Beete Jukes. *Athenaeum,* February 13, 1858: 203.

Richards, Evelleen. "Huxley and Woman's Place in Science: The 'Woman Question' and the Control of Victorian Anthropology." In *History, Humanity, and Evolution,* ed. James R. Moore, 253–84. Cambridge: Cambridge University Press, 1989.

Rodden, Judith. "The Development of the Three-Age system: Archaeology's First Paradigm." In *Towards a History of Archaeology.* See Daniel, 1981.

Rogers, Emma, ed. *The Life and Letters of William Barton Rogers.* 2 vols. Boston: Houghton Mifflin, 1896.

Rogers, Henry Darwin. "The Reputed Traces of Primeval Man." *Blackwood's* 88 (1860): 422–38.

Rudwick, Martin J. S. "Charles Darwin in London: The Interaction of Public and Private Science." *Isis* 73 (1982): 186–206.

———. "Charles Lyell's Dream of a Statistical Paleontology." *Paleontology* 21 (1978): 225–44.

———. "Cognitive Styles in Geology." In *Essays in the Sociology of Perception,* edited by Mary Douglas, 219–41. London: Routledge, 1982.

———. "The Foundation of the Geological Society of London: Its Scheme for Cooperative Research and Its Struggle for Independence." *British Journal for the History of Science* 1 (1963): 325–55.

———. *The Great Devonian Controversy: The Shaping of Knowledge among Gentlemanly Specialists.* Chicago: University of Chicago Press, 1985.

———. "Hutton and Werner Compared: George Greenough's Geological Tour of Scotland in 1805" *British Journal for the History of Science* 1 (1863): 117–35.

———. "International Arenas of Geological Debate in the early Nineteenth Century." *Earth Sciences History* 5 (1986): 152–58.

———. "Levels of Disagreement in the Sedgwick-Murchison Controversy." *Quarterly Journal of the Geological Society of London* 132 (1976): 373–75.

———. "The Shape and Meaning of Earth History." In *God and Nature,* 296–321. See Lindberg and Numbers, 1986.

Rupke, Nicolaas A. *The Great Chain of History.* Oxford: Clarendon Press, 1983.

Salter, J. W. Letter. *Geologist* 1 (1958): 301–3.

Sarjeant, William A. S. *Geologists and the History of Geology: An International Bibliography from the Origins to 1978.* 5 vols. New York: Arno, 1980.

Schneer, Cecil J. "The Rise of Historical Geology in the Seventeenth Century." *Isis* 45 (1954): 256–68.

"The Science of Anthropology," *Eclectic Review* 15 (1868): 471–79.

Secord, James A. *Controversy in Victorian Geology: The Cambrian-Silurian Dispute.* Princeton, NJ: Princeton University Press, 1986.

———. "J. W. Salter: The Rise and Fall of a Victorian Paleontological Career." In *From Linnaeus to Darwin,* edited by Alwyne Wheeler and J. Price, 61–75. London: Society for the History of Natural History, 1985.

———. "King of Siluria: Roderick Murchison and the Imperial Theme in Nineteenth Century Geology," *Victorian Studies* 25 (1982): 413–42.

———. "The Geological Survey of Great Britain as a Research School, 1839–55." *History of Science* 14 (1986): 223–75.

Sedgwick, Adam. "Address of the President." *Proceedings of the Geological Society of London* 1 (1820): 281–316.

Silverberg, Robert. *The Mound Builders of Ancient America: The Archaeology of a Myth.* Greenwich, NY: New York Graphic Society, 1968.

Smart, T. Wake. "Account of Some Ancient British Antiquities Discovered a Few Years Ago in Kent's Cavern, Near Torquay, Devon." *Journal of the British Archaeological Association* 2 (1845): 171–74.

[Smith, William H.] "Wilson's *Prehistoric Man.*" *Blackwood's* 93 (1863), 525–44.

Spencer, Frank. *Piltdown: A Scientific Forgery.* Oxford: Oxford University Press, 1990.

———. "Prologue to a Scientific Forgery: The British Eolithic Movement from Abbeville to Piltdown." In *Bones, Bodies, Behavior,* edited by George W. Stocking, Jr. (Madison: University of Wisconsin Press, 1988), 84–116.

Stanley, William Owen. "On the Remains of Ancient Circular Habitations in Holyhead Island, with Notes on Associated Relics by Albert Way." *Archaeological Journal* 24 (1867): 229–64.

Stocking, George W., Jr. "The Ethnographer's Magic: Fieldwork in British Anthropology from Tylor to Malinowski." In *Observers Observed,* 70–120. Madison: University of Wisconsin Press, 1983.

———. *Victorian Anthropology*. New York: Free Press, 1987.

Swayne, G. C. "The Value and Charm of Antiquarian Study." *The Antiquary* 1 (1880): 3–5.

Taylor, E. Reginald. "The Humours of Archaeology." *Journal of the British Archaeological Association*, n.s. 38 (1932): 183–234.

Taylor, John. "The Hand of Man in Kirkdale Cavern." *Macmillan's* 6 (1862): 386–98.

Todhunter, Isaac. *William Whewell, D.D.: An Account of His Writings, with Selections from His Literary and Scientific Correspondence*. 2 vols. London: MacMillan, 1876.

Tomline, George. "Inaugural Address at the Suffolk Congress." *Journal of the British Archaeological Association* 21 (1865): 1–4.

Trautman, Thomas R. *Lewis Henry Morgan and the Invention of Kinship*. Berkeley and Los Angeles: University of California Press, 1987.

Trevelyan, W. C. "Works of Art in the Drift." *Athenaeum*, December 31, 1859: 889.

Trevor-Roper, Hugh. *The Romantic Movement and the Study of History*. London, 1969.

Trigger, Bruce G. *A History of Archaeological Thought*. Cambridge: Cambridge University Press, 1989.

Tylor, Alfred. "On the Discovery of Supposed Human Remains in the Tool-Bearing Drift of Moulin-Quignon." *Anthropological Review* 1 (1863): 166–68.

Tylor, Edward Burnett. "Lake Dwellings." *Quarterly Review* 125 (1868): 418–40.

Van Keuren, David K. "From Natural History to Social Science: Disciplinary Development and Redefinition in British Anthropology, 1860–1910." In *The Estate of Social Knowledge*, edited by JoAnne Brown and David Van Keuren, 45–66. Baltimore: Johns Hopkins University Press, 1991.

———. *Human Science in Victorian Britain: Anthropology in Institutional and Disciplinary Formation, 1863–1908*. Ann Arbor, MI: University Microfilms International, 1982.

Vivian, Edward. "On the Earliest Traces of Human Remains in Kent's Cavern." *Annual Report of the British Association for the Advancement of Science* 26 (1856): 119–23.

W. C. "The Antiquity of Man." *British and Foreign Evangelical Review* 16 (1867): 383–400.

Walford, Edward. Preface. *The Antiquary* 1 (1880): i–iv.

Wallace, Alfred Russel. "The Origin of the Human Races and the Antiquity of Man Deduced from the Theory of Natural Selection," *Anthropological Review* 2 (1864): clviii–clxx.

[Way, Albert.] "Introduction." *Archaeological Journal* 1 (1844): 1–6.

Webb, R. K. *Modern England.* 2d ed. New York: Harper, 1980.

[Wedgwood, Julia.] "Sir Charles Lyell on the Antiquity of Man." *Macmillan's Magazine* 7 (1863): 476–87.

[Westmacott, Richard]. "Introduction." *Archaeological Journal* 7 (1850): 1–7.

White, Malcolm, "Does Scripture Settle the Antiquity of Man." *British and Foreign Evangelical Review* 21 (1869): 128–37.

Whitley, Nicholas. *Are The Flint Implements from the Drift Authentic?* London: Longman's, 1865.

———. "Brixham Cave and Its Testimony to the Antiquity of Man, Examined." *Journal of the Transactions of the Victoria Institute* 8 (1874–75): 211–24.

———. "The Paleolithic Age Examined." *Journal of the Transactions of the Victoria Institute* 8 (1874–75): 3–23.

———. "True and False Flint Implements." *Popular Science Review* 8 (1869): 30–38.

Wilkinson, Gardner. "Remains of Man in Caves." *Athenaeum,* June 9, 1860: 791.

Willis, Delta. *The Hominid Gang: Behind the Scenes in the Search for Human Origins.* New York: Viking, 1989.

Wilson, Leonard G., ed. *Charles Lyell's Scientific Journals on the Species Question.* New Haven, CT: Yale University Press, 1970.

[Wood, H. H.] "Lake Dwellings of Switzerland." *Contemporary Review* 4 (1867): 380–94.

Woodbridge, Kenneth. *Landscape and Antiquity.* Oxford: Clarendon Press, 1970.

Woodward, Arthur Smith. "William Boyd Dawkins." *Dictionary of National Biography,* supplement, *1922–30,* 250–51.

Woodward, Horace B. *A History of the Geological Society of London.* London: Geological Society, 1907.

———. "Robert Alfred Cloyne Godwin-Austen." *Geological Magazine,* n.s. decade 3, vol. 2 (1885): 1–10.

Wright, Thomas. *The Celt, the Roman, and the Saxon.* London: Hall, 1852.

————. *The Celt, the Roman, and the Saxon.* 2d ed. London: Hall, 1861.

————. *The Celt, the Roman, and the Saxon.* 3d ed. London: Trübner, 1875.

————. "Flint Implements in the Drift." *Athenaeum,* June 18, 1859: 809.

————. "On the Progress and Present Condition of Archaeological Science." *Journal of the British Archaeological Association* 22 (1866): 64–84.

————. "On the Remains of a Primitive People in the South-East Corner of Yorkshire." In *Essays on Archaeological Subjects.* 2 vols. 1:1–21 London, 1860.

————. "Works of Art in the Drift." *Athenaeum,* July 9, 1859: 51–52.

Wrottesley, Lord. "Presidential Address." *Annual Report of the British Association for the Advancement of Science* 30 (1860): lv–lxxv.

York, archbishop of. "Inaugural Address to the Annual Meeting of the Royal Archaeological Institute." *Archaeological Journal* 24 (1867): 83–91.

Young, John R. *Modern Scepticism in Relation to Modern Science.* London: Saunders, Otley, 1865.

Young, Robert M. *Darwin's Metaphor.* Cambridge: Cambridge University Press, 1985.

INDEX

Abbeville, 62, 168. *See also* Somme Valley

Adams, W. H. Davenport, 232, 237

Africa, 228

Age, of earliest humans: absolute vs. relative, 151–52; "agnostic" view of, 163–65; diversity of views on, 156–71; as divisive issue, 148, 154, 156–57, 172–73, 182–83; "Lyellian" view of, 152–54, 157–61; "Prestwichian" view of, 152–54, 161–65; and radiometric dating, 227–28; "traditionalist" view of, 165–69

Amateur scientists: in geology, 54–58, 76, 84; in historical archaeology, 15, 21–28, 215; in prehistoric archaeology, 215–16

Anderson, John, 123–25, 232, 235

Ansted, David T., 149, 232, 235

Anthropology: boundaries of, 183; and human antiquity, 226–27; and prehistoric archaeology, 186–87, 197, 201–2, 205, 211; and prehistory, 205. *See also* Evolution, cultural; Evolution, human; Fossils, human; Paleoanthropology

Archaeologia, 18, 28, 31, 229

Archaeological Institute, 22, 26, 130

Archaeological Journal, 28, 31, 230

Archaeological journals, 34, 36–37, 39, 184, 190, 229–31. See also *Archaeologia; Archaeological Journal; Journal of the British Archaeological Association*

Archaeological societies: founding of, 15, 21–26; goals of, 26–29; membership of, 26; and prehistoric archaeology, 190; research programs of, 33–35. *See also* Archaeological Institute; British Archaeological Association; Historical archaeology; Society of Antiquaries, London

Archaeology: boundaries of, 3–4, 59, 142–43, 183–87, 191–92, 204–5; in 18th century, 16–20; and geology, 184, 204–5; impact of human antiquity on, 184–221, 226–27; Victorian popularity of, 15, 22–26, 190–91. *See also* Historical archaeology; Prehistoric archaeology

Argyll, Duke of, 154, 164, 180, 232, 236

Artifacts: as art objects, 18–20; ecclesiastical, 25; interest in